THE REALLY PRACTICAL GUIDE TO PRIMARY SCIENCE

Second Edition

Carol Holland and John Rowan

Stanley Thornes (Publishers) Ltd

First edition published in 1992 by:
Stanley Thornes (Publishers) Ltd
Ellenborough House
Wellington Street
CHELTENHAM GL50 1YW
England

Second edition 1996

96 97 98 99 00 / 10 9 8 7 6 5 4 3 2 1

A catalogue of this book is available from the British Library.

ISBN 0 7487 2616 0

Printed and bound in Great Britain at The Bath Press, Avon.

Contents

Introduction and acknowledgements   v
How to use this book   vi

Section A: Teaching primary science

Chapter 1: Making sense of National Curriculum science   1

What is science?   1
The framework of National Curriculum science   1

Chapter 2: Looking at science in your school   3

What is a scheme of work?   3
Reviewing present practice   3

Chapter 3: Teaching and learning issues   6

An overview of the issues   6
Classroom organisation and management   6
Equal opportunities   9
Communication   11
Recording findings   12
Links with other curriculum areas   13
Questioning   14
Safety   15
Resources   16
Record-keeping   18
Assessment   23

Chapter 4: Planning your scheme of work   27

Planning for structure, continuity and progression   27
Choosing a model for your scheme of work   28
Planning at classroom level   30
Using a weekly planning and assessment sheet   33

Chapter 5: Evaluation   37

Evaluating activities   37

Evaluating topics 38
Evaluating the scheme of work 38
Conclusion 39

Section B: Topics for classroom use

Chapter 6: Infant topics 41

Water 41
Me and others 45
Toys 49
Look up, look down, look around 52
Celebrations 56
Moving on land 59
Moving in water and in the air 62

Chapter 7: Lower junior topics 65

Clothes 65
Changes 67
Communications 70
Transport 73
Supermarket 75
On a desert island 78

Chapter 8: Upper junior topics 81

All around our school 81
Lights, sound, action! 84
Buildings and builders 87
Energy 90
Our world and beyond 93
Healthy living 96

Chapter 9: Sources of information 99

Chapter 10: Photocopiable sheets 100

Introduction

The National Curriculum is an 'entitlement curriculum' for all children of all abilities between the ages of 5 and 16. The implementation of the National Curriculum offers a unique opportunity to plan and deliver an effective programme of learning placed within the broader context of schools' overall curricular aims and policies. Science, as one of the core subjects, holds a key place.

The major aim of this book is to support the effective teaching of science in key stages 1 and 2, showing how science enriches the curriculum and how it can be implemented effectively, building on the good practice already evident in our primary schools.

I believe that this book can help you to ensure that your children experience the broad and balanced science curriculum which is their entitlement. I hope that it will also help you to enjoy science and pass that enjoyment on to the children.

Acknowledgements

Thanks are due to all those colleagues in Manchester and St Helens who provided support and advice during the development of this book.

Special thanks are due to Keith Johnson whose help and support has been invaluable.

Thanks are also due to my husband Keith for his patience and understanding which made the project possible; and to my son Neil for his patience, and his culinary and typing skills which lightened my load considerably.

Carol Holland

I would like to thank my family for ignoring me during the updating for this second edition.

John Rowan

How to use this book

This book is a *practical* guide to planning and teaching science in primary schools. Because of the emphasis of the National Curriculum the book is structured around its requirements, as outlined in the revised document *Science in the National Curriculum* (1995). However, even if you are not closely following the National Curriculum you will find the book valuable for planning, teaching and assessing primary science in a structured and effective way.

This book is in two parts:

Section A: Teaching primary science

This covers, in a clear and straightforward way, all the issues that are involved in planning, teaching and assessing primary science. Chapter 1 gives you a clear analysis of *Science in the National Curriculum* and what its implications are for you. Successive chapters take you step-by-step through all the elements of effective practice in your class and school. A number of valuable photocopiable planning sheets are also provided and you will find these at the back of the book.

This section is intended as a guide to better practice. Whatever the state of primary science teaching in your school, you will probably find that a study of these chapters will enable you to plan and teach more effectively, and to correct imbalances in your present teaching.

Section B: Topics for classroom use

This section contains a comprehensive ideas bank for primary science. For each two-year group there are six units of work (plus a unit for reception children) that can be used either as a complete off-the-peg scheme of work or as a flexible ideas bank. There is effectively one topic for each term of the school year, complete with flow charts, core ideas, extension work and National Curriculum links.

You do not have to read Section A to make use of the Section B ideas bank. It is a freestanding resource which will provide you with a huge bank of structured ideas, even if you are totally confident about the issues outlined in Section A.

1

MAKING SENSE OF NATIONAL CURRICULUM SCIENCE

What is science?

Science is essentially a way of thinking and working.

It must include:

- The development of basic skills
- The fostering of positive attitudes
- The development of scientific concepts.

Children learn best through active involvement in learning experiences. Through this involvement they develop ideas which help them to make sense of the world around them. Their knowledge and understanding of science will both develop progressively through their schooling. Work in key stages 1 and 2 provides the foundation for later years. By channelling the children's natural curiosity into scientific investigation we can help them to acquire strategies to develop more formal and complex concepts.

Good science education is evident where children are:

- Involved in practical investigations which build upon their previous experiences and interests
- Developing the skills of hypothesising, observing, classifying, measuring, recording and inferring
- Cooperating with others in planning, investigating and communicating
- Gaining scientific knowledge and understanding

- Reflecting on their work
- Taking responsibility for investigations
- Responding to stimuli provided for them
- Relating their work to the real world
- Demonstrating attitudes such as self-discipline, curiosity, perseverance and open-mindedness.

The framework of National Curriculum science

National Curriculum science is described through five programmes of study (PoS): Foundation Science and Science 1, 2, 3 and 4. These set out what pupils should be taught. The attainment targets (ATs) set out the expected standards of pupils' performance at six levels (1, 2, 3, 4, 5 and 6).

The PoS for key stages 1 and 2 are further sub-divided as shown on page 2.

Understanding Foundation Science

Foundation Science contains the essential knowledge, skills and understanding that will need to be applied across the other four PoS in order to study them successfully.

Understanding Science 1

Science 1 is concerned with the basic pro-

1

Foundation Science

Systematic enquiry

Science in everyday life

The nature of scientific ideas

Communication

Health and safety

Science 1

Experimental and Investigative Science

Planning experimental work

Obtaining evidence

Considering evidence

Science 2

Life Processes and Living Things

Life processes

Humans as organisms

Green plants as organisms

Variation and classification

Living things in their environment

Science 3

Materials and their Properties

Grouping materials (and classifying, KS2 only)

Changing materials

Separating mixtures of materials (KS 2 only)

Science 4

Physical Processes

Electricity

Forces and motion

Light and sound

The Earth and beyond (KS 2 only)

cesses and skills of science. Contexts derived from Science 2, 3 and 4 should be used to teach children about experimental and investigative methods. Children should also be given the opportunity, as part of normal science activities, to carry out whole investigations themselves. Investigations are activities in which children ask questions, test ideas, predictions or hypotheses and draw conclusions.

Few children use the results from their investigations to come to their conclusions. This skill, in particular, requires careful teaching. The difficulty of an investigation depends on the concepts which underlie it. It is best to start from the child's understanding of the scientific ideas involved. Because of planning, assessing and resourcing implications, it is suggested that a maximum of one investigation per child per half-term be undertaken.

Understanding Science 2, 3 and 4

Science 2, 3 and 4 are concerned with knowledge and understanding of science. Science 2 is concerned with the biological aspects; Science 3 with the chemical and Science 4 the physical. Emphasis is placed on children being 'taught' these sections. Care should be taken to relate the teaching to pupils' existing knowledge and to use familiar contexts.

2

LOOKING AT SCIENCE IN YOUR SCHOOL

What is a scheme of work?

In order to meet the requirements outlined by the programmes of study and the attainment targets, every school will need to develop a structured framework for learning. This framework is best described as a scheme of work, and the way in which it fits into the overall school planning is shown by the diagram below.

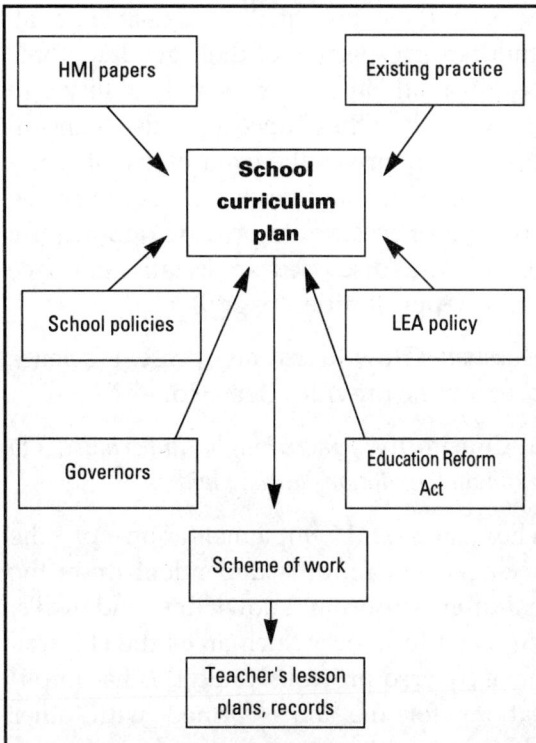

```
┌─────────────────┐        ┌──────────────────┐
│   HMI papers    │        │ Existing practice │
└─────────────────┘        └──────────────────┘
          │                         │
          ▼                         ▼
              ┌──────────────┐
              │    School    │
              │  curriculum  │
              │     plan     │
              └──────────────┘
          ▲                         ▲
          │                         │
┌─────────────────┐        ┌──────────────────┐
│ School policies │        │    LEA policy    │
└─────────────────┘        └──────────────────┘

┌─────────────────┐        ┌──────────────────┐
│   Governors     │        │ Education Reform  │
│                 │        │       Act         │
└─────────────────┘        └──────────────────┘
                   │
                   ▼
           ┌──────────────┐
           │Scheme of work │
           └──────────────┘
                   │
                   ▼
           ┌──────────────┐
           │Teacher's lesson│
           │plans, records │
           └──────────────┘
```

The school's scheme of work is a way of setting out a structure for the learning experiences which children should have during their time in school. It must be sufficiently structured to ensure coverage of the programmes of study in an appropriate way, but it also needs to be sufficiently flexible to allow you as the teacher to respond to the needs and interests of your children.

A scheme of work will have more detail in it than the school policy statement for science. It does not, however, pre-empt the formulation of the policy statement which should describe the broad aims of the nature and scope of the science to be taught. The scheme must never be as rigid as a syllabus, however, as this implies a fixed and inflexible list of content to be taught and, as such, cannot respond to the needs of your class and your children.

Reviewing present practice

Ten principles for a scheme of work

It is important that, before a scheme of work is produced, the whole staff discuss the style, nature and principles of the scheme. If the scheme of work is to encourage cooperative classroom learning it is essential to involve the whole staff.

The DES booklet *Science 5-16: A Statement of Policy* (1985) suggested the ten principles outlined below as the basis for developing a scheme of work. These issues must be addressed and the following questions considered.

1 Breadth *All pupils should be introduced to the main concepts from the whole range of science... and to a range of skills and processes.*

In the past, early science has been approached with the emphasis on nature study and many children have not had the necessary introduction to the 'physics' aspect. The programmes of study for key stages 1 and 2 provide a framework for the concepts and skills but it is essential that we review the present practice in our own classrooms.

Question: How well am I covering the main areas of science?

2 Balance *All ... should achieve a balance between the acquisition of scientific knowledge and the practice of scientific method.*

The PoS outline the requirements of both the process and the content. Science 1 contains details of the skill development necessary as children progress through the key stages. Science 2, 3 and 4 provide the contexts in which children can practise the skills.

Question: Am I providing a balance of the necessary skills and fostering the development of them in appropriate contexts?

3 Relevance *Science education should draw extensively on the everyday experiences of the children.*

Children do not learn in 'isolated pockets'. By starting from their own interests and experiences we can ensure that their learning is appropriate to their stage of development. The children will also feel a sense of ownership of the programme of work; for example, a topic on toys which stems from the children discussing their own toys and investigating how they work will allow them to experience Science 4 in a way which will seem more relevant to them.

Question: How do I ensure relevance now?

4 Differentiation *The intellectual and practical demands... should be suited to the abilities of the pupils... for all pupils.*

Within any class there are children working at several different levels. It is necessary that we address this when planning specific activities within a topic and ensure that all levels are catered for.

Question: How do I cater for the most and least able, and also for the majority of my children?

5 Equal opportunities *... should give genuinely equal curricular opportunities... actively seek ways of exciting the interest of girls.*

Science must be equally accessible to all children irrespective of their gender, ethnic background and intellectual ability. To deny children the chance to realise their full potential deprives the community of many valuable skills and resources. Some of the strategies necessary to provide opportunity for all are discussed in detail later (see Equal Opportunities, page 9).

Question: How does my present science programme provide science for all?

6 Continuity *Increasingly important... to build on foundations already laid.*

The successful implementation of the science curriculum is dependent upon the children acquiring knowledge and skills. We need to know which areas the children have covered previously and this has implications for the links formed with other classes, departments and schools.

Question: What curricular links do I have now with other classes/departments ?

7 Progression *Courses should be designed to give progressively deeper understanding and greater competence.*

Revisiting areas must never entail repetition of what has gone before. Each visit will normally be at a higher level, will demand a deeper understanding of knowledge and a more sophisticated use of the necessary processes.

Question: In what ways can I ensure that I am building on previous work?

8 Links across the curriculum *Should link science ... with the development of their language and mathematical competence.*

It is difficult to envisage science without measurement or without communication, and it provides a vehicle for children to practise mathematical, technological and language skills. Many science topics also provide a forum in which to address cross-curricular themes, skills and dimensions.

Question: How can I use science as a vehicle for other areas of the curriculum?

9 Teaching methods *Science is a practical subject and should be taught in a way which emphasises practical, investigative and problem-solving activities.*

Children learn most productively when they are actively involved in practical investigations. They will only develop the skills outlined in Science 1 if we provide the opportunities for first hand experiential learning. This has obvious implications for the way in which we organise the learning environment and this issue is addressed in more detail later.

Question: How can I organise my classroom and my teaching style to ensure that children learn productively?

10 Assessment *Progress should be assessed ... in ways which allow all pupils to show what they can do, rather than what they cannot.*

Assessment helps us to plan the next steps in children's learning and should be seen as part of their ongoing activity within the classroom. Assessment activities should be incorporated in the teaching programme and not simply be an end-of-topic activity.

Question: How can I ensure that I have the opportunity to assess my children?

By using these ten principles and addressing the questions in the light of current practice it will be possible to highlight the areas of greatest concern and develop an action plan suitable for your children and your school. In forming an action plan which will lead to the development of a working scheme, you will need to address what to teach, how and when to teach it and ways in which to record the outcomes. These issues are discussed in the next two sections. It is important to recognise that this is a whole-school issue and that although ultimately one person or group of people will take on the responsibility of putting the scheme together, any discussions on the formulation of the scheme must include all members of staff. In separate infant and junior schools there must be liaison to ensure that we are building on previous experience.

3

TEACHING AND LEARNING ISSUES

As well as the actual content of the scheme of work there are a number of issues to be considered. As one of the main aims of a scheme of work is to provide a structured framework for learning, it must include other aspects of teaching and learning. These are illustrated in the diagram below and are discussed in the following sections of the chapter.

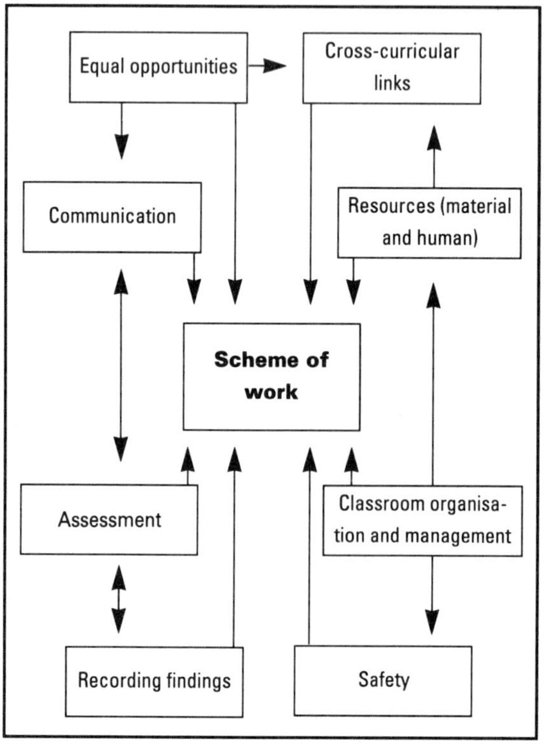

Classroom organisation and management

Ways of organising science work

Science is a practical activity and this has implications for your classroom organisation. There are several different ways of organising science, using a whole-class or a group approach, which are set out below. You will need to select the approach or combination of approaches which best suits the needs of your children, which is viable in your classroom and which suits your style of teaching.

In each example the following code applies:

S = Children working on a science activity

C = Children working on other activities

T = Teacher.

The arrow sizes on each diagram reflect the amount of teacher intervention required.

→ A high intervention activity where you need to spend a fair amount of time with the children, for discussion, for safety reasons or because you are assessing the children.

→ An activity which requires only intermediate intervention on your part.

------► A low intervention activity

which the children are able to complete with little or no intervention from you.

Approach 1 All the class working on the same science activity.

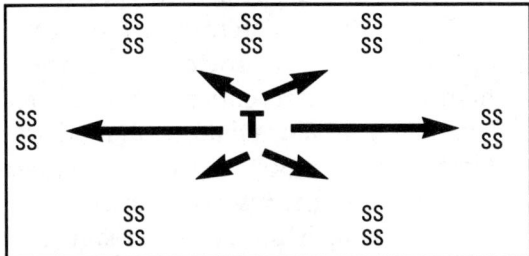

Advantages

- First hand, practical experiences for all
- Easy planning
- Develops cooperation.

Disadvantages

- One activity may not be suitable for whole class
- Resourcing could be costly
- Teacher intervention – all groups could need help at the same time.

Approach 2 A whole-class activity.

There are several organisational stages in a whole-class activity:

1 Introduction to the whole class

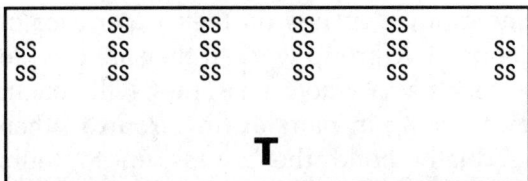

2 Groups working on different aspects

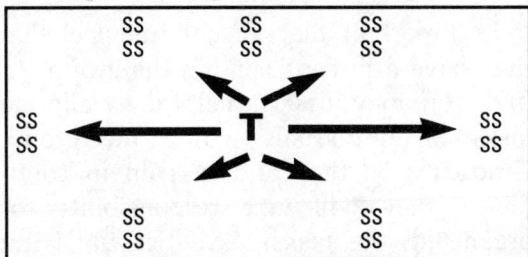

3 The whole class together for testing and discussion.

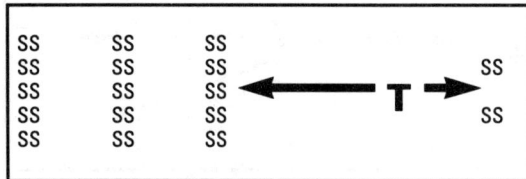

Advantages and disadvantages

- Similar to approach 1, but this one is useful if you are working on a problem-solving activity such as designing and making a paper plane. It makes the testing a much safer proposition and your results are likely to be more accurate. It will also help the children to appreciate the importance of listening to the results of others and incorporating others' outcomes in their own work.

Approach 3 All groups working on a circus of science activities.

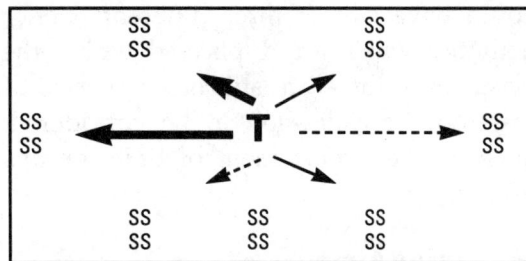

Advantages

- High on motivation
- Small amounts of a variety of equipment
- Teacher intervention manageable if the activities are planned carefully.

Disadvantages

- Timing could be difficult
- Speed and sequence problems
- Could be demanding on the teacher.

Approach 4 One or two groups working on science activity.

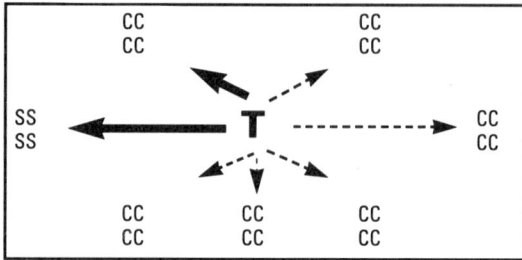

```
CC          CC
CC          CC

              ↖  ⇡ →
SS  ←——————  T  ------→  CC
SS                        CC
              ↙  ⇣ ↘

CC      CC      CC
CC      CC      CC
```

Advantages

- Teacher intervention more manageable
- Works well in integrated day
- Easy to resource
- Easier to match activity to individual needs.

Disadvantages

- Needs careful planning
- Has timing implications
- Reporting back may be difficult.

Having chosen an approach or adapted one of them to meet your own needs (for example, in the third outlined approach you could have two or three different science activities which are duplicated within the classroom), there are still other aspects of classroom organisation to be considered, such as the management of time, groups and resources.

Managing time

Teacher time You will need to have realistic goals which:

- Allow non-contact time for planning, talking to parents, and so on
- Allocate the amount of time to be spent on the science aspect of the theme (remembering all your other commitments)
- Allow time for visiting speakers and visits.

Children's time To use children's time most effectively you need to:

- Train the children to manage their own time (using tick lists, sheets, time limits, etc.) but remember to be sensitive to the children's individual work pace
- Provide a mixture of activities. Here it is useful to think in terms of the traffic-lights model and provide a balance of high teacher-intervention activities (red – certainly no more than two), moderate teacher-intervention activities (amber) and low teacher-intervention activities (green – which are, in the main, self-sustaining/independent activities.) It is important that the children realise the significance of the traffic-lights model, to which they should respond readily. In my own classroom, having identified the red, amber and green activities, I would tell the children my expectation of their work programme, i.e. they had to complete two red, one amber and two green activities during the day
- Use written instructions to assist independence, develop reading skills and encourage cooperation, such as readers helping non-readers.

Managing groups

How many children in a group? The ideal number of children working together in one group depends on their experience of cooperative group work. If they are used to working as a whole class, they will benefit by working in pairs at first. You can then gradually build the groups up to fours (which is the optimum number for most activities).

It is important that all children feel that they have a purpose within the group. At first you may find it helpful to allocate tasks to individuals within the group. Gradually, as the children gain in confidence, they will take responsibility for organising the tasks. Activities involving sorting and classifying in pairs are good for building up group cooperation.

How many groups working on science activities? This will depend on the availability of resources and the nature of the activities. Some activities demand a high level of teacher intervention and it is better to balance such an activity with others on which the children can work profitably with little attention from you. For example, in a food topic one group could be baking with you, while others are making observational drawings of fruits and vegetables, conducting taste tests, making models of favourite meals, conducting surveys of favourite foods or writing out recipes.

Where will the groups work? There are several aspects which need to be considered here. First there is the safety aspect. If children need space to work (for example, if they are testing which toy vehicle travels furthest or investigating jet propulsion with balloons) they will need a quiet area away from the usual classroom traffic. It will also encourage in them a sense of responsibility if they feel that they are trusted to work in the corridor or hall. (I have often found that the more I trusted my children to work without direct supervision, the more they responded by staying on task.)

Another aspect is the noise level of an activity. Activities which involve noise, such as sawing and hammering, can be distracting to others and so they are best kept to the sides or corners of the room.

The activity itself has implications on where the children will work. As you well know, when children are carrying water from one end of the classroom to another, spillage is inevitable. If you are not lucky enough to have a water tray or sink in your classroom try to ensure that activities involving water are near to the door.

Organising resources

General equipment such as scissors, sticky tape, pencils, paper and glue should ideally be stored in a central area within the room. It is essential that children know where such items are kept and have reasonable access to them if you are to foster a sense of responsibility in looking after equipment and promote independent learning.

It is also helpful to encourage the children to be responsible for collecting materials, and for putting them away after they have finished! (I know that sometimes it is easier and quicker to clear away yourself, but perseverance works wonders.)

Timing and sequencing of activities

Inevitably some activities take longer to complete than others so it is essential that children know what to do when they have finished. It is also necessary to remember that there may be a logical order to the activities, and children may become frustrated or lose interest if they cannot find an answer to a question because they have not worked through a preliminary activity.

Equal opportunities

A well documented feature of our educational system is the underachievement in science of children from ethnic minorities and of girls. This level of underachievement deprives individuals within these groups of the chance to realise their full potential whilst depriving society of valuable skills and resources.

The media bombard us with a stereotypical image of scientists. If you were to ask children in your class to draw or describe a scientist, it is likely they would portray a white, middle-aged, male 'professor' clad in a white coat. Many children reject a career in science from an early age as they feel that they do not fit the stereoptypical image. As teachers we are in a strong position to change this. The introduction of National Curriculum science – an 'entitlement' curriculum for all children of all ages – is a powerful weapon as it will broaden children's experiences of science throughout their compulsory schooling and help to foster a positive attitude to science.

Children learn best by doing, and the skills which they develop will help to foster important attitudes, such as cooperating with others, open-mindedness and perseverance, which are positive attributes for everyone. Science can be an excellent vehicle for promoting equal opportunities if we translate the principle of 'science for all' into positive classroom practice.

There are many strategies which can be used to try to achieve this aim.

Cross-cultural links

By making cross-cultural links integral to our chosen themes we can provide cultural breadth for all children, and foster positive attitudes by valuing a variety of cultures and traditions. For example we can:

- Choose topics which are rich in cross-cultural links, such as food, celebrations, clothes and ourselves (Sc 2, 3, 4)
- Ensure that alternative languages, lifestyles, customs and scientific achievements of other cultures are incorporated into displays and published materials
- Arrange visits to and from religious and cultural centres to share customs, practices, festivals, music and dancing (Sc 3 and 4)
- Borrow artefacts such as multicultural clothes, musical instruments, ornaments (Sc 4)

- Visit shops and markets in multi-ethnic areas to show the wide variety of available foods, fruits, vegetables and spices (Sc 2 and 3)
- Invite parents, grandparents and friends from minority groups into school to share their experiences (Sc 2)
- Remember to focus on the similarities between people – we have more similarities than differences (Sc 2).

Girls and primary science

Females in our society are seriously under-represented in scientific careers. This appears to start at an early age: APU tests at age 11 found 'marked differences between boys and girls in using some measuring instruments and applying physics concepts'.

By the age of 11 children's views of science can be firmly entrenched, which means a change in attitude must be presented throughout the primary years.

Here are some suggestions for fostering a positive attitude in girls towards science from an early age:

- Provide models of women as scientists and technicians
- Encourage girls to take an active exploring role in groups rather than a passive, scribing one
- Ensure that women teachers show themselves as competent role models, at ease with tools, construction kits, computers, and so on
- Invite male and female visitors in non-traditional occupations into school
- Teach all children the scientific principles underlying traditional female occupations such as cookery, textiles, child care
- Provide extra support and more opportunities for girls to work with other girls

using construction kits, machines, electricity, and so on

- Encourage girls to develop their spatial awareness by ensuring experiences with large apparatus, model-making
- Use gender-neutral language ('made' not 'man-made'; 'people' not 'mankind')
- Check published materials for sexual stereotyping. (Are women shown in non-traditional occupations? Do pictures show girls taking an active role?)

Communication

The processes and content of science would be meaningless without communication. The essential communication skills and the various techniques which children should develop throughout their science education are outlined in the communication strand of the Foundation Science PoS. There are also references to communication skills in Science 1, 2, 3 and 4. The attainment target level descriptions set out appropriate types and range of performance that pupils working at a particular level should demonstrate. By bringing together this information and some specific examples, it is possible to see progression as children work through the key stages.

During key stage 1, all children should have opportunities to:

- Name
- Describe
- Present information in speech and writing
- Communicate what happened during their work
- Make a record of observations and measurements
- Use drawings, labels, tables and bar charts to present results

- Say what they have found out from their work.

During key stage 2, all children should have opportunities to:

- Use appropriate scientific vocabulary
- Use standard measures and SI units
- Use a wide range of methods to record and present information in an appropriate and systematic manner, for example: diagrams, drawings, tables, charts and line graphs
- Use their recorded information to point out and interpret patterns or trends in their data
- Draw conclusions consistent with the evidence and begin to relate these to scientific knowledge and understanding.

There are many ways in which we can encourage purposeful discussion in the classroom.

Interaction between children

If we provide opportunities for the children to work in pairs we encourage vocabulary enrichment as well as helping them to suggest ideas. It is only by talking freely that children are able to shape their own ideas and absorb the patterns which they find.

We must also encourage children to use their results purposefully, for example asking the children to find out information from a bar chart. In order to ensure that discussion between children is purposeful we need to monitor that we are offering opportunities for them to:

- Follow intructions
- Listen to others
- Contribute to discussion
- Offer ideas
- Report in sequence.

Discussion between children and children, and children and teacher

At the end of a session it is useful to give children the opportunity to report back to others in the class. This encourages the development of listening skills, helping the children to value other people's ideas, as well as giving them more insight into their own investigations.

Reporting back to others requires specific skills on the part of the children and also much patience from you, but nevertheless it is worth the effort. It is useful if you encourage the audience to ask questions. This helps the children to clarify their thinking and the way they present their findings. Again, results take time!

Whole-class discussion

This is useful either as a starting point to stimulate ideas or at the end of a session to draw threads together. The most important issue to remember here is to keep the sessions short (between ten and fifteen minutes at the most).

Communication to the whole school and to parents

The school assembly is a useful forum for the encouragement of verbal communication skills. As well as talking about investigations and models which they have produced, children can comment on slides of themselves working through activities.

Recording findings

It is essential that the children record their findings. It will help them to develop communication skills and it will also help to provide evidence for assessment purposes.

Children inevitably see recording as an outcome of an activity. (Does this remark, overhead on the return journey from a visit to a farm, sound familiar? 'I bet I know what we'll be doing tomorrow – she'll have us writing about the animals and drawing pictures. We always do …')

It is essential that children are encouraged to jot down pointers during their investigations. This will provide ongoing records of long-term observations, such as the growth of a plant, time-lines or a diary of the weather. It will also act as an *aide-memoire* in problem-solving activities if they sketch prototype models showing amendments or improvements.

Recording after investigations helps the children to draw conclusions, make inferences and suggest hypotheses which lead to further investigation.

It is essential that children not only record their findings in ways which are appropriate to them, but also that they develop an understanding of why they need to record results.

There are many ways in which children can record their results. It is important that they experience as many ways as possible:

- Talking/reporting
- Model making
- Written reports
- Diagrams and pictures
- Tape recording
- Video recording
- Using the computer
- Drama
- Poetry
- PE
- Dance
- Photographs
- Tables, bar charts, line graphs. (This must be considered as a three-stage process. In

the first stage children will complete ready prepared tables. Later they will begin to draw up their own simple tables. In the third stage they will be ready to select the most appropriate type of chart for their results and then design it.)

- The group report. (Here the group of children working together will decide how to present a joint record of their work. They might design a poster or present a cooperative display of tables, written reports, pictures, photographs, etc. in which each group member has made a contribution. This will help children to recognise and utilise the skills and talents of others.)

The above list is by no means exhaustive. The most important thing to remember is to vary the ways in which we ask children to record – sometimes a finished model is perfectly adequate.

Links with other curriculum areas

Science has many links with other areas of the curriculum. Other areas can be used to communicate scientific findings: PE, drama and movement lessons are excellent vehicles for such a purpose. In art and craft sessions, when children are mixing colours and experimenting with different media, they are developing scientific skills and knowledge in addition to developing creative and aesthetic skills. When they are communicating findings verbally or in written form the children are experiencing English AT1: Speaking and listening, and AT3: Writing.

It would be difficult to imagine science without measurement or handling data and so the links with mathematics are strong. For example, in many scientific investigations

children will experience mathematics AT1: Using and applying mathematics. Throughout their scientific education children will also be meeting SoAs from AT2: Number and algebra, AT3: Shape, space and measures, and AT4: Handling data.

In many primary classrooms children will be meeting criteria laid out in the technology document, although they may not at first glance all be involved in working on a 'technological' activity. Consider your own classroom. Depending upon the age-ranges of the children you may have various activities taking place. For example, in a nursery or infant classroom you may have children playing in the sand and water tray, others may be acting out a situation in the structured play area. If you teach older children, some may be conducting a survey, others working on a mini-enterprise.

Whatever the age-group, some of your class might be reading in the quiet area, or using construction kits, or making models from junk materials. In any of the above activities children are meeting the criteria of the programmes of study for technology:

> *"Pupils should be taught to develop their design and technology capability through combining their designing and making skills with knowledge and understanding in order to design and make products."*
>
> Design and Technology in the National Curriculum,
> DES, 1995

In many scientific themes children will be meeting technology criteria. Here are some examples:

1 *Health and safety.* The ability to recognise and assess hazards and risks, and to take action to control them, are common to both science and technology.

2 *Working with materials and components.* How the working characteristics of materials

relate to the ways in which they are used. Working with electrical and mechanical components. In these situations children are meeting both science and technology criteria.

3 *Developing and communicating ideas.* Technology criteria are met whenever children draw their models, design machines, take measurements or make sketches.

4 *Focused practical tasks and investigations.* Children develop and practise particular skills and knowledge, working independently and in teams.

In a science topic on movement, a tried and tested activity is to design and make a wheeled vehicle. An extension to this could be to add tread to the wheels and see if this affects the way the vehicle travels. In this one activity the children are experiencing the following ATs:

Mathematics AT1: Using and applying mathematics, AT2: Number and algebra, AT4: Handling data

English AT1: Speaking and listening

Science AT1: Experimental and investigative science, AT3: Materials and their properties, AT4: Physical processes

Technology AT1: Designing, AT2: Making

In addition to the many curriculum areas with which science has traditional links, many scientific activities and investigations will help children to experience cross-curricular skills, themes and dimensions such as citizenship, health education, industrial awareness, environmental education and equal opportunities.

Questioning

Questioning should encourage the children

to express their thoughts and ideas. Thinking aloud helps them to develop and explore their understanding of the world around them. A major aim of questioning is to promote purposeful activity, so open questions are far and away preferable to closed questions (those which have only one acceptable answer, often either 'yes' or 'no'). Many children are reluctant to answer closed questions because of their fear of giving the 'wrong' response. Open questions do not have a single 'correct' answer and tend to elicit a more positive response.

This is also the case with person-centred questions; for example, to ask, 'Why do *you* think ...?' is not likely to be threatening to the child and will also give you more detail about the child's reasoning ability.

When questioning children there are other aspects to consider.

- Tone of voice. It is important to sound encouraging and friendly – the questions shouldn't sound like interrogation.
- Valuing the child's response is of paramount importance and a most powerful tool. This can be achieved by using praise, tone of voice, using the child's name and responding positively with smiles and nods.
- Repeating the child's answer. This ensures that all the other children can hear the response, and is also useful for emphasising a particular section of the response.

Asking open and pupil-centred questions

We ask questions for a variety of purposes. For example, in an investigation into finding the most suitable material to make an ice cube last longer, the questions can enable children to:

- Review their ideas – 'What can you tell me about using the newspaper to insulate the ice cube?'
- Observe in more detail – 'What can you tell me about the different materials?'
- Pursue investigations more purposefully – 'How much better was the polythene than the tissue paper in making the ice cube last?'
- Stop and rethink – 'How can we make sure that our test is fair?'
- Justify their ideas and actions – 'Why did you make all the pieces of material the same size?'
- Encourage self-criticism – 'If you were going to do this investigation again, what do you think you would change?'
- Demonstrate their understanding of a particular scientific concept by applying their findings to a practical situation – 'What material would you use now if you wanted to keep something cold? For example, what would you use to wrap a block of ice-cream?'

Safety

At all times the safety of the children is of prime importance and it is one of your key responsibilities as class teacher. Science activities which involve children taking responsibility for planning their work need careful supervision. As teachers acting *in loco parentis* you must ensure that the children work in a way which is safe. In some situations that might mean suggesting alternative strategies to the children.

An essential part of the learning process is to encourage children to become independent learners. To fulfil this you must encourage the children to take responsibility for the collecting together and clearing away of materials which they may use within an investigation, so remember to leave enough time for this. It is important to instil into the children a responsibility for keeping their working area clear and uncluttered during their investigations. By choosing beforehand the required materials and equipment we can reduce any problems which might occur.

A safety checklist

- Provide water-based glues rather than rubber-based or solvent ones. PVA is clean, non-toxic and washes out if accidentally spilled on clothes
- Plastic beakers are easily obtainable, cheaper and much safer than glass ones
- Spirit thermometers must always be used. All 30 cm-long thermometers are clearly labelled and are easily read by young children. Mercury thermometers must not be used as the liquid mercury from a broken thermometer is a serious health hazard
- Naked flames, such as candles or nightlights, must only be used under strict supervision. Chidren with long hair should have it tied back and loose cuffs should be rolled up. Ensure that activities involving naked flames take place on a sand tray and that a fire bucket is nearby
- Do not use hot glue guns. If you wish to use a glue gun, low temperature glue guns are available and much safer.
- Do not use chemical salts such as copper sulphate and cobalt chloride. These are poisonous if swallowed. The scientific value of making chemical gardens or growing crystals using these salts is questionable
- Never use very hot or near boiling water (50°C max.). Supervise or pour out hot water yourself and keep hot kettles out of the children's reach
- Warn the children never to taste or inhale

15

substances without supervision

- Ensure that children wash their hands before and after any cooking or tasting activity. Make sure that all utensils are clean
- Warn children never to eat any leaves or berries they may collect
- If you have pets in the classroom remind the children of the need for strict hygiene. Treat any bites with scrupulous care.

Encourage the children to take responsibility for safety. A list such as the following will help to make the working areas safer.

---

1 Keep work areas tidy. Put away any equipment you have finished with.

2 Clean any spillages right away.

3 Keep bags in the cloakroom, well away from work areas.

4 Make sure that everyone in your group is working safely.

5 If you are not sure what to do, ask your teacher.

Resources

The teacher

As a teacher your role is complex and difficult to define. It depends upon many things: your confidence, the previous experiences of your children, and the nature of learning. Within the classroom you will take on many roles. For example, during the course of a day you may be:

An enabler, guiding the children towards areas of research

A manager, coordinating the activities and managing material and human resources

A presenter, offering activities, giving information and clarifying ideas

An adviser, listening, suggesting alternatives, offering encouragement

An observer, studying the children, monitoring their progress and giving feedback

An evaluator, assessing the children's progress and monitoring the activities.

Often the best advice you can give the children is directing them towards appropriate resources to help them find their own solutions. This will mean that you will have to consider the materials available in the classroom.

Material resources

Specialist apparatus is not necessary but children should have easy access to a range of simply equipment in order to carry out investigative activities. The list opposite, though not exhaustive, is a typical example of basic equipment which most schools have.

The following are excellent sources of free materials:

- Shoe shops and supermarkets for boxes and cardboard
- Dairies for yoghurt pots
- Electrical and engineering firms for wire and magnets
- Timber yards for wood offcuts.

Parents as a resource

Parents are a valuable resource that is continuously under-used. In addition to providing materials and helping on visits they are invaluable in the classroom. Initially they may only want to work with their own children but, as their confidence increases, they will be willing to work with other groups.

The ground rules for parental involvement in the classroom should be laid down at the outset by you and the parents. It is important that parents and teachers work as partners and that the parents' skills are fully utilised.

A questionnaire sent home asking for details of people willing to commit a regular period of time, specifying suitable times and outlining specific skills which parents feel they have to offer to the school is a good idea. Many parents have computing, electronic, scientific and artistic skills which are far superior to your own and which would enhance the children's education. Yet many of these skills are under-used simply because you are not aware of them.

It is essential that parents understand their role within the classroom, that they are aware of the need to let the children pursue their own lines of enquiry and also realise the importance of open-ended questioning. A parents' evening which shows parents the ways in which children learn through active

---

A materials checklist

To enhance observation:
hand lenses
'lem' viewer
mini-spectors
midi-spectors
stereo (binocular) microscope

To develop measuring skills:
metre rules
centimetre rules
measuring jugs
measuring cylinders
height charts
tape measures
tocker timers
Transan timers
water timers
stop clock/watch
bathroom scales
balances
electronic weighing machine
slotted masses
droppers
thermosticks
spirit thermometers
spring balances
Newton meters
centicubes

To develop investigating skills:
torches
plastic mirrors
acetate sheets
concave/convex mirrors
lenses
kaleidoscope
stethoscope

plastic tubing
filter paper
blotting paper
funnels
magnets
batteries (1.5v, 4.5v)
bulbs (1.5v, 2.5v)
bulb holders
simple motors
buzzers
switches
plastic-covered wire
screwdrivers
crocodile clips
cotton reels
cogs
pulleys
beads
marbles
springs (Slinky)
carpet tiles
slope
battery-operated toys
wind-up toys
balloons/balloon pump
balls
elastic
paper fasteners
plastic guttering

To develop designing and making skills:
junior hacksaws
dowel
wood (offcuts)
square section wood
balsa wood

hammer
hand drill
sandpaper
masking tape
wheels – card, wood, plastic
straws
Plasticine®
clay
plaster of Paris
Mod-Roc
nails
screws
corks
pipe-cleaners
string
syringes
Corriflute
materials (fabrics)
junk materials (boxes, etc.)
propellers
Capsela
Duplo®
Lego®
Lego Technic®
Meccano®
Teko®
gears of many different types

To encourage exploration of the environment:
pooters
Petri dishes (plastic)
plastic tanks
plastic bowls
buckets
sieve
natural materials, rocks, fossils

---

involvement will not only increase their confidence to work in the classroom but will also give them an insight into the curriculum.

To have confident parents working alongside you in the classroom will make your role more manageable as you will be able to plan activities around parental skills and they will provide you with more opportunities to assess groups within your class.

Record-keeping

The purpose of record-keeping is to improve the effectiveness of schooling by ensuring a better match between the abilities of children and their learning experiences. An agreed school marking/record-keeping policy is essential.

Keeping records of the children's experiences is important for several reasons:

- In order to deliver the National Curriculum successfully, it is essential that all children revisit each part of the programme of study more than once in each key stage
- It will be important to know exactly which areas of the programme of study have been experienced by your children in the past. Only then will you be able to plan a programme of learning which ensures adequate progression in both skill development and concept development
- Valuable curriculum time can be wasted by unnecessary repetition of activities. It is important to revisit a skill area or concept but it is boring to revisit exactly the same activity.

Records will enable you to:

- Monitor the breadth, balance and relevance of planned activities

- Address continuity and progression in teaching and learning
- See at a glance the areas which have been covered and those still to be covered and allow for purposeful revisiting without unnecessary repetition.

It follows that records should give evidence of:

- Topics and activities experienced
- Skills experienced
- Concepts developed.

Record-keeping systems fall into two main groups:

- Records of the experience of a whole class
- Records of an individual child's attainment.

Class records

To provide a complete summary of the work covered by a particular class, several records are needed. The first of these is a copy of the topic web that was used to plan the half term's work, such as the one opposite. Topic webs will be more or less useful depending on the amount of detail that they show and this varies from school to school. However, a topic web does set the context of the children's learning and is thus useful as an overview.

The next record is a suggested topic planning and assessment sheet (see page 20). Schools may wish to adapt this to use for each half-termly topic. It can provide such information as:

- The name of the topic
- The learning objectives from the programme of study/scheme of work, i.e. knowledge, understanding and skills
- The activities experienced by the children
- The methods of recording used
- Assessment opportunities

- The skills from other subject areas, such as English and mathematics, which can be practised while working on science activities, may also be recorded
- An evaluation of the activity which will be useful in future planning.

The completed sheet on page 20 shows how the topic planning and assessment sheet can be used as a planning tool as well as for recording. The topic in this example is a Y6 topic on Communications.

1 Having chosen the topic, draw up a topic web like the one below. The main areas to be covered are identified by consulting your school scheme of work and the relevant programme of study.

2 The next step is to select appropriate learning objectives and activities for the children (having checked class records to ensure that revisiting does not mean repetition of an activity) and to enter the details in the boxes on the planning and assessment sheet. Objectives should be as concise as possible, focusing on the knowledge, understanding and skills the children should acquire.

It is not necessary to put something in every box, contrived links within a topic are seldom helpful.

3 Identify the children's probable method of recording. It is valuable to return to this column at the analysis stage as the children may find unexpected ways of recording and communicating their findings.

4 Where specific resources are needed these should be recorded. It may be a piece of apparatus that needs to be borrowed, for example a video, computer disk or textbooks.

5 The assessment column encourages you to look for built-in assessment opportunities (rather than bolting on separate tasks or tests at a later stage). A useful shorthand to indicate the method of assessment uses the letters O, P, Q or S:

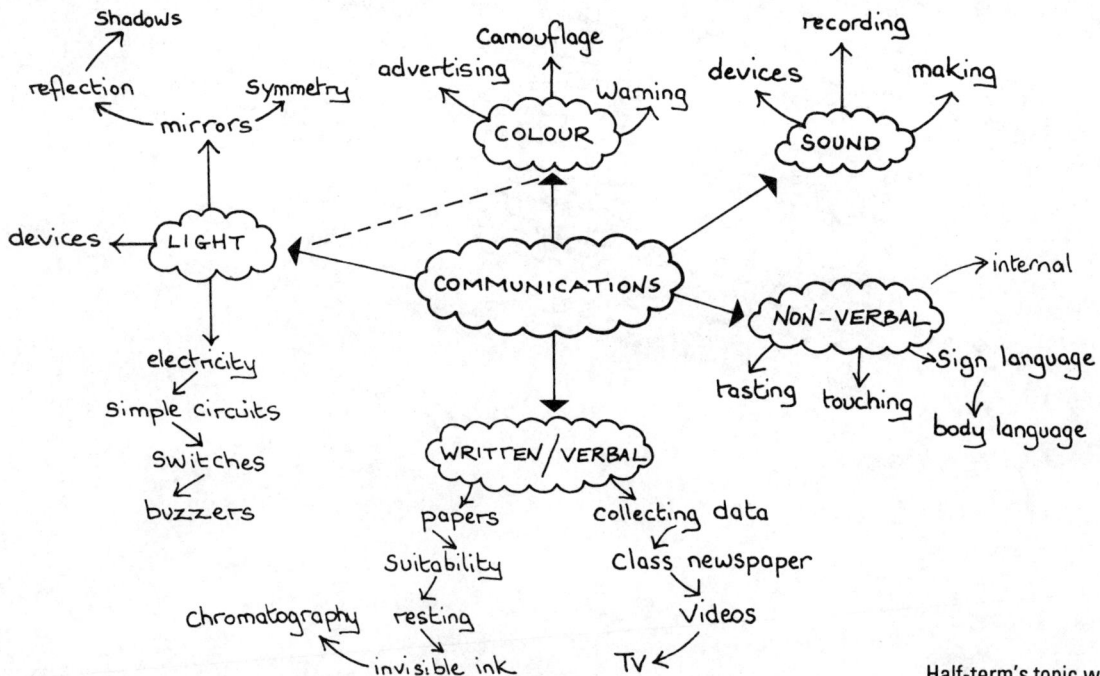

Half-term's topic web

O = assessment by observation of the children

P = assessment by product, for example the writing, drawing, artefacts produced by the children

Q = assessment by questioning (this could include written tests)

S = self assessment

The columns completed so far will provide a

Topic planning and assessment sheet

Class......Y6a......

Area of study/topic......Communication......

Term......Autumn......

Area of study/topic	Learning objective (from PoS)	Activities	Method of recording	Specific resources	Assessment (O, P, Q, S)	E
Science	1b: to be able to use focused exploration and investigation to acquire scientific knowledge, understanding and skills	Via Sc2 investigation	Written report	As below		
Foundation		Via Sc2 investigation	Written report	As below		
Science 1	1a: to turn ideas suggested to them, and their own ideas, into a form that can be investigated; 1d: to carry out a fair test or comparison		Written report	Stopwatches, cocktail stick/blu tac, pulse meters	P/Q	
Science 2	2e: to know the effect on pulse rate of exercise (internal communication)	To carry out an investigation in which the children try to establish the link between exercise, pulse rate and breathing rate	Chart	Range of objects of a similar size/shape made from wood, metal, plastic, etc. Feely bag. Recyclable materials	O	
Science 3	1a: to be able to compare everyday materials on the basis of their properties	Sort materials by touch and grade them according to own criteria. Design and make a poster for a blind/partially sighted person	Poster	Electronics equipment	P/S	
Science 4	1b: to know how switches can be used to control electrical devices; 1c: to know and use ways of varying the current in a circuit to make bulbs brighter and dimmer; 3g: to know that vibration from sound sources can travel through a variety of materials, eg metals, glass, air to the ear	Make a flashing sign which could be used to advertise either a product or a shop. Design and make the best string telephone	Finished product / Written report	Different containers. String	O/Q/P / P	

basis for advance planning of the topic.

6 The final column allows for a simple evaluation of the activity as a learning experience for the children. It is important for schools to develop their own criteria for this evaluation.

Information from the topic planning and assessment sheet can be transferred to the class record sheet outlined below. This is only for key stage 1, though a similar one could be used for key stage 2. There are copies of these at the end of the book. The sheet has been drawn out so that each PoS statement has one box.

The children's experiences can be recorded by marking the relevant boxes as indicated below.

After the 1st visit

After a re-visit

Foundation and Science 1 will be revisited on many occasions. The number of visits can be recorded as:

The sheet provides a summative record of the work covered by the class and serves as a useful reminder of the need to widen children's experiences. It can aid transition to the next class and provide supplementary information for use in moderating a class's work with others.

Schools will need to decide the exact format and quantity of recorded information, and during assessment 'best-fit' with the level descriptions is advised. It is still as important as ever to remove the 'fresh-start each year' syndrome. School record-keeping systems must allow subsequent teachers to carry on where the previous teacher left off. It may well be that in some schools a highly structured scheme of work reduces the need for whole-class record-keeping to a 'deficit model' in which only omissions from that year's work are recorded. In others, the assessment system may hold the key. Whatever the system, teachers will need to learn to trust each other's professional judgements. Standardisation meetings involving the whole staff are invalauble in achieving this.

Science class record sheet (Key stage 1)

Class .... to .... 19............

Foundation Science	1	a	b	c	d
	2	a	b	c	
	3	a			
	4	a	b		
	5	a	b		
Science 1 Experimental and investigative science	1	a	b	c	
	2	a	b	c	
	3	a	b	c	d	e	f
Science 2 Life processes and living things	1	a	b		
	2	a	b	c	d	e	f
	3	a	b	c	
	4	a	b		
	5	a	b		
Science 3 Materials and their properties	1	a	b	c	d	e
	2	a	b		
Science 4 Physical processes	1	a	b	c	
	2	a	b	c	d
	3	a	b	c	d	e

Topics covered

Year group	Autumn	Spring	Summer	Teacher
Nursery				
R				
1				
2				

21

Used jointly, the topic web, topic planning and assessment sheet and the class record sheet will provide valuable information. This could be:

- Formative (helping to plan appropriate activities)
- Evaluative (showing any under coverage of the PoS)
- Informative (to parents, governors, LEA)
- Transitive (to pass on from teacher to teacher, school to school)
- Summative (a record of the curriculum experienced).

Individual records

Individual records are a necessary part of building up each child's overall profile/record of achievement.

To be useful, the records should provide a succinct summary of:

- The work covered
- The skills gained
- The concepts understood
- The child's perception of his/her learning.

The individual records can then act as a focus for discussion with the child, parents and colleagues (particularly on transition).

It is suggested that work in a child's book/topic folder, etc. can form the main record of an individual's progress. Work can be suitably annotated by the teacher to provide immediate feedback to the child and to form a record of the child's competence in a particular area. A useful way of recording ephemeral evidence that may otherwise be lost in the day-to-day classroom happenings is to make a note, as it happens, on a 'Post-it' and stick it on the child's work, or on your mark book, for your later reference.

At appropriate intervals (decided by the

whole staff) a summative record of individual children's progress can be made. It may be useful to confirm this by a test/task of some sort. This is particularly useful to confirm that the knowledge base has been learned. Tests/tasks can take a wide variety of forms, pencil and paper being only one of many assessments which are discussed in more detail in the section on assessment (page 23). The date on which the child displayed competence is useful, as there will be instances where a child is unable to do so at a later date.

Self-assessment is an important aspect of the record-keeping, though it is related more to the activities than the outcomes. Involving children in the assessment of their own work is a useful way of confirming and enhancing progress. Children who understand why they are doing something and what they have to do to succeed invariably work better. There should be opportunities for children to set their own targets. Schools may wish to link this in with the production of a record of achievement or portfolio of work.

For some children it may be useful to have some ready printed annotation slips with suitable comments such as:

'This piece of work shows that I can ...'
'I like this work because ...'
'This piece of work shows I need more practice at ...'

The primary science pro forma shown on page 23 could be used to involve the child in self-assessment. The topic/activity column would usually contain information transferred from the class topic planning and assessment sheet.

The next two colums can contain code letters entered by the child. The code would be agreed by discussion. For example:

Code 1:

A = I enjoyed this work
B = I liked some bits
C = I didn't enjoy it

Code 2:

E = I found it quite easy
M = It wasn't too hard
H = I found this work hard

Younger children could use symbols:

Even simple feedback from the children gives useful information about their understanding and enjoyment of the work.

Primary science pro forma			
Name .... Class .... Date			
Topic/activity	Code 1 2	Child's comment	Teacher's comment

Children will need some initial help and encouragement with self-assessment.

Genuine open encouragement is needed if children are to persuaded to declare their thoughts and feelings about their learning. Most children will need help to be positive and constructive about their learning rather than negative and downgrading. With young children you can paraphrase their oral responses to a few key questions. Your own comments will be more useful, especially at transition to another class if the staff as a whole agree on the key aspects of learning to be considered and commented on.

Assessment

The assessment of children's development in science is a matter of concern to all teachers. Key issues include the time required, the skills needed, the nature of external tests/tasks (STAs) and the possible distortion of the teaching programme.

These concerns are natural. In many ways they mirror the concerns of secondary teachers just before the advent of the GCSE exams. However, after the first run-through of the GCSE, secondary teachers generally felt confident about their new ways of assessing children's progress and HMI reported significant improvements in the quality of children's learning.

In July 1993, the Dearing Interim Report stated:

> *"One of the issues that has arisen during consultation is that when a national test has taken place, little or no weighting is currently given to teacher assessment in the tested attainment targets. This is to under-value soundly based, moderated teacher assessment. My recommendation for all key stages is that the national*

test and teacher assessment ratings should be shown separately in all forms of reporting and in school prospectuses."

The national tests at the end of each key stage and teacher assessment provide different information. Teacher assessment is wide ranging and formative, the tests are summative providing a snapshot at a particular time. The tests are unable to cover all the attainment targets, experimental and investigative science being left to the teacher.

The test results should not be seen as a way of confirming or invalidating teacher assessment results. When using the level descriptions to look at pupil performance, teachers need to judge which description best fits. There is no expectation of an exact match.

Assessment must be seen as part of the teaching process and form an integral part of the learning programme as outlined below.

If while planning the children's programme of learning you have considered assessment opportunities (see Record-keeping, page 18) you are better able to form an objective view of the children's progress.

Collecting and recording evidence

In order to gain and maintain an ongoing picture of the child's developments there needs to be a developing body of evidence which reflects the child's activities and learning outcomes. Crucially, this will illustrate what the child knows, understands and can do. This evidence may include:

● Your notes; an ongoing record, possibly a mark book which has one page for each child. This can be completed as the need arises and could be about achievement, more experience needed or simply day-to-day information. At the end of each week you could read the comments, taking note of those children with few comments and perhaps targeting them for the following week

● A developing portfolio of selected pieces of work (bearing in mind that one piece of work can provide evidence of several ATs

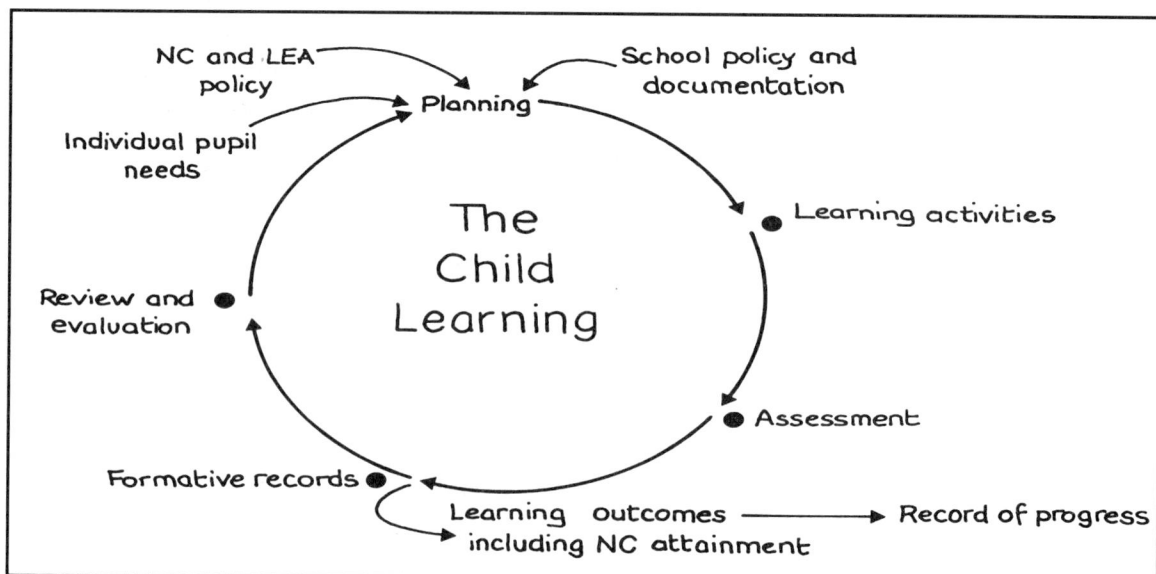

NC and LEA policy

School policy and documentation

Individual pupil needs

Planning

The Child Learning

Learning activities

Review and evaluation

Assessment

Formative records

Learning outcomes including NC attainment

Record of progress

in several curriculum areas) perhaps as part of a record of achievement

- Opportunities for pupil and parental contributions. Children could choose work to go in their portfolio and could be involved in the updating by choosing what to remove. Parents could be involved by being encouraged to add comments to children's work.
- Individual/group/class discussions to review objectives. These can be formal or informal at the beginning or end of the sessions and will provide insight into the children's perceptions of their learning
- Performance on problem-solving activities, investigations and collaborative tasks. By including an egg-race type of activity it is possible to observe how the children react to a challenge and to group interaction, as well as demonstrating that learning is fun
- Records of oral work/practical demonstrations. These could be part of a class assembly or form a section of an open afternoon for parents
- Using photographs as a record. This will involve the children in an explanation of the purpose and context.

In undertaking assessment it is important to recognise that activities involve learning processes as well as products, and a variety of methods, including observation, must be developed.

Observing children

It is essential to bear in mind a number of points before undertaking any observation of a practical activity.

- Be clear about the aim and purpose of the assessment. Does the activity really provide an appropriate assessment opportunity? (See topic planning and assessment sheet.)

- Decide on a limited number of skills /concepts to assess. (See topic planning and assessment sheet.)
- Focus your thinking on the skill(s)/ concept(s) and how they will be exhibited, before you start. What evidence will you be looking for?
- Scrutinise and internalise the words on your record sheet as much as possible beforehand so that you can focus on the activity rather than the sheet.
  Do not try to assess too many children in any one session, certainly no more than one group of three or four children.
- Try a 'dummy-run' on another group to help you to highlight the key features of the activity.
- Aim to spend an appropriate length of time on the observation. One or two minutes will be too short for you to fully understand the activity in which the children are engaging whilst fifteen minutes is too long, given the needs of the rest of the class. Five minutes will give you significantly more information especially if you return for another observation a few minutes later. As a guideline, think of the time you would need to spend in listening to the group reading.
- You may learn more about a child's skills if you act as an uninvolved observer, rather than trying to be the teacher, facilitator and tutor simultaneously.
- Organise the rest of the class with self-directed and absorbing tasks as far as you can. Remember you are only talking about limited times.
- Record the skills you observed immediately after the observation, aiming as far as possible to be unprejudiced by a child's (in)ability in another curriculum area or by a child's misbehaviour.

Asking yourself the following questions may help to improve and refine your observation skills:

- What did the children actually do?
- What did I expect the children to do?
- What did the children learn?
- What did I learn?
- What did I plan next for the children?
- What did the children say about their learning?

The golden rule is to set yourself practical, realistic and achievable targets.

Informal and formal teacher assessments are vital elements of the assessment procedure. They must be continuous if they are to provide a true picture of what a child can do. You need to sample and target groups of children at different times and in different situations. This will enable you to assess a class of children, over a half-term or a term's work, within a structured framework. Remember, if assessment informs subsequent work, you are on the right track.

4

PLANNING YOUR SCHEME OF WORK

Planning for structure, continuity and progression

In the past most teachers chose science topics which covered approximately half a term's work. Topics were usually chosen which met some or all of the following:

- They were child-centred
- They matched the needs and interests of children and teacher
- They covered the main areas of science.

To meet the requirements of the Science National Curriculum it is necessary to add further criteria:

- Children should be taught the programmes of study, ensuring they visit and revisit each area within each key stage
- More emphasis must be placed on continuity and progression so it is essential that the programmes of learning are planned by the whole staff.

A structured learning programme should be planned which ensures continuity and progression, while retaining the essential principles of good primary practice. The programme or model chosen must ensure that the children receive the broad, balanced curriculum laid out in the programme(s) of study for their appropriate key stage(s).

The spiral curriculum

A scheme of work for science should describe a spiral curriculum. That is, it should ensure that a child visits and revisits (several times if possible) each concept area of the curriculum. Each visit will normally be at a higher level, demanding a deeper understanding of that area of knowledge and a more sophisticated use of process skills.

You may wish to picture the spiral curriculum as a staircase spiralling upwards to greater comprehension, or as a route spiralling downwards to greater depths of understanding. For children with specific learning difficulties it may be that, initially, the turns of the spiral are closer together.

Another way of picturing the curriculum is to imagine it spiralling outwards from the child, as the child's conceptual understanding of the world around widens and as his or her interests broaden.

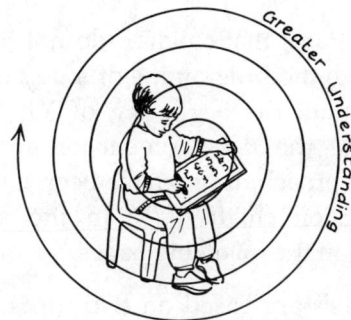

You will need to decide how many turns of the spiral should occur during a child's time in your school. Should there be one complete turn of the spiral during each school year – i.e. the whole of the programme of study for the relevant key stage visited each school year? Or would it be more realistic to consider two turns of the spiral in the two-plus years of infant education where the body of knowledge is smaller? This would then change in the junior department to two turns of the spiral during the four years. That would mean that children would cover the programme of study twice in the junior department – once in Y3 and Y4, and once again but at a higher level in Y5 and Y6. This method would be more realistic in ensuring adequate coverage of the body of knowledge in the programme of study for key stage 2. The timing of the national tests (STAs) will also have an influence.

Choosing a model for your scheme of work

There are several models which will ensure that the children cover all the ATs and it will be necessary for all the staff to decide on the model or combination of models which best suits the school's need. It would be unrealistic to leave the choice of topics to individual teachers as it would be difficult to ensure continuity and progression, difficult to monitor and might not cover all the attainment targets.

The law states that children do not have to commence the programme of study for key stage 1 until the beginning of Y1, but to encourage the development of a whole-school approach, topics have been included for reception children (R) in the models outlined on the following pages.

The models are based on two turns of the spiral in the infant department and a further two turns in the juniors. The topics shown are for illustration only although they do cover all the attainment targets within the turns of the spiral. Each topic covers half a term's work.

Some of the advantages and disadvantages of each model have been outlined, but a full staff discussion would highlight pros and cons specific to individual schools.

Model 1

Fixed topics using a whole-school approach (see opposite). All the age-groups study the same topic every half term.

Advantages

- Ensures depth
- Assists progression
- Easy to plan and monitor
- Continuity clear
- Encourages staff interaction.

Disadvantages

- Heavy demand on resources
- Lacks spontaneity
- Individual interests ignored.

Model 2

Topics which are based on individual programmes of study (see opposite).

Advantages

- Ensures coverage of programmes of study.

Disadvantages

- May not engage interest of children
- Not child-centred
- National Curriculum as a minimum entitlement
- Primary practice at its worst
- No cross-curricular links.

Model 1

Model 1: Fixed topics, a whole school approach

	Autumn		Spring		Summer	
	Ourselves	Celebrations	Movement	Communications	Materials	Change
R						
Y1						
Y2						
Y3						

Model 2

Model 2: Topics based on programmes of study

	Autumn		Spring		Summer	
Ro					Sc 2 (1),(2)	Sc 2 (1)
Y1	Sc 2 (4),(3)	Sc 3 (1)	Sc 3 (2)	Sc 4 (3)	Sc 4 (1)	Sc 4 (2)
Y2	Sc 4 (2)	Sc 4 (1),(2)	Sc 2 (1),(2)	Sc 2 (4)	SATs	Sc 4 (1)
Y3	Sc 2 (1)	Sc 2 (2),(4)	Sc 3 (1),(3)	Sc 2 (3)	Sc 3 (2)	Sc 4 (1)
Y4	Sc 4 (2)	Sc 4 (3)	Sc 4 (4)	Sc 2 (2)	Sc 4 (2)	Sc 4 (1)
Y5	Sc 2 (1),(2)	Sc 2 (4)	Sc 2 (3)	Sc 3 (2)	Sc 3 (1)	Sc 3 (3)
Y6	Sc 4 (1)	Sc 4 (2)	Sc 4 (3)	Sc 4 (4)	SATs	Transition Work

Model 3

Fixed topic on a class-by-class approach (see opposite). Each class studies individual topics.

Advantages

- More balanced approach
- Can ensure breadth and depth
- Staff aware of what precedes and follows.

Disadvantages

- Appears rigid
- Interests of children and teachers may be ignored.

Model 4

Fixed and open topics (see opposite). Each class will study one fixed topic per term, leaving the other half term for a free choice topic.

Advantages

- Allows adequate coverage of PoS if topics chosen with care
- Some opportunities for spontaneity
- Few resource implications
- Can respond to children's interests.

Disadvantages

- Needs great care to ensure coverage of all PoS
- Needs frequent re-appraisal.

Having chosen an appropriate model it will be necessary to consider the sequence of topics to be covered, bearing in mind:

- The age and previous experiences of the children
- The season (for example, minibeasts are difficult to find in January, and the most reasonable time to ask children to look at the night sky is in the winter when it gets dark early)

- The need to provide a balance of the main areas of science over a year.

Once the sequence of topics to be covered has been completed, the next stage will be to check that the topics ensure coverage of all the relevant programmes of study.

Model 5

Fixed and open topics (see page 32)

If at this stage there appeared to be under-coverage of some PoS the topics could be rearranged to correct this.

The next step in the programme is to plan each individual topic in detail and to consider what the children are actually going to do. Examples of ways to do this are given in Section B (page 40).

Planning at classroom level

There are four steps to take when planning a topic in detail in order to translate ideas into practical activities which are well-matched to the children's abilities.

Step 1. Choose a topic

Probably the topic will already have been identified in your school scheme of work. However, it is necessary to consider several factors before planning in fine detail:

- The children's previous experiences
- The children's interests
- Equal opportunities (gender equality, cultural diversity and special needs of all kinds)
- Cross-curricular skills (such as communication, numeracy, study, problem solving, personal and social, and information technology)
- Cross-curricular themes (economic and

Model 3

Model 3: Fixed topic, class approach

	Autumn		Spring		Summer	
R					Water	Sounds
Y1	All about me	Autumn	Sky and space	Play-grounds	Mini-beasts	People who help us
Y2	Colour	Clothes	Toys	Weather	Homes and homes	Food
Y3	Using our senses	Celebrations	Weather	In the kitchen	Our school	Holes
	Materials	Communications	Structures	Light	Toys and games	Hook up and down
	Wood	Light and dark	Sky and space	Sound	Plants	Flight
	Around school	Moving on water	Time	Materials	Transport	Change

Model 4

Model 4: Fixed and open topics

	Autumn		Spring		Summer	
R					Water	
Y1	Me and others		Toys		Look up, look down, look around	
Y2	Celebrations		Moving on land		SATs	Moving in water and in the air
Y3	Clothes		Changes		Communications	
Y4	Transport		Supermarket		On a desert island	
Y5	All around our school		Lights, sound, action!		Buildings and builders	
Y6	Energy		Our world and beyond		Healthy living	

Model 5: Fixed and open topics

	Autumn		Spring		Summer	
R					Water	
					Sc1, 3(1) (3), 4 (1) (3)	
Y1	Me and others		Toys		Look up, look down, look around	
	Sc1, 2(1) (2)		Sc1, 4 (1) (2)		Sc1, 2 (3) (4), 3(1), 4 (4)	
Y2	Celebrations		Moving on Land		SATs	Moving in water and in the air
	Sc1, 3(1)(2), 4 (1) (3)(4)		Sc 1, 3(1), 4 (1) (3)			Sc 1, 3 (1), 4 (1) (3)

industrial understanding, careers, health education, citizenship and environmental education). You can find more help on equal opportunities, cross-curricular skills and themes in *Curriculum Guidance 3: The Whole Curriculum* (The National Curriculum Council, 1990)

- The duration of the topic
- The time of the year (matching the topic to the season, and to the children's stage of development in the school year).

Step 2. Produce a topic web

Produce a topic web by:

- Brainstorming ideas
- Referring to the relevant paragraphs from the programme of study
- Grouping the ideas into areas (e.g. focusing on curriculum areas).

Step 3. Complete a planning and assessment sheet

This can either be the topic planning and assessment sheet described on page 18 or the weekly planning and assessment sheet – which is shown on page 34.

Whichever is used, the result will be a detailed plan of the learning activities. You will need to consider:

- An appropriate starting point
- The resource implications and the resource needs (human and material)
- The identification of equal opportunities aspects and other awareness opportunities (i.e. cross-curricular themes)
- The highlighting of skills to be experienced
- The matching of the activities to the children's needs and abilities. (Can the

activities be differentiated to stretch the more able children? Will all children be able to achieve success?)

- Breadth, balance and relevance in the range of activities
- The identification of assessment opportunities
- The identification of methods of recording and communicating
- The identification of the relevant programmes of study
- Timetabling implications (e.g. use of the hall, assembly, swimming, playground duties, human resources when parents come in to help).

Step 4. Make a daily plan of work

This should indicate:

- The allocation of learning experiences to all groups (this will depend upon your style of classroom organisation)
- The differentiation of activities where necessary
- Likely teacher intervention (Is the programme manageable? Is there a balance of high, moderate and low intervention activities?)
- Strategies for recording, evaluation and assessment
- The ongoing activities: mathematics scheme, English scheme, RE, etc.

Using a weekly planning and assessment sheet

You may find the weekly planning and assessment sheet, shown on the next page, the best way of organising your detailed planning. The criteria used for the development of the outlined sheet were to create a planning tool which:

- Takes ideas from a topic web and/or topic

planning and assessment sheet, and then encourages planning in detail for a block of time (perhaps one or two weeks)

- Identifies specific knowledge, understanding and skills
- Assists in the matching of activities appropriate to the children's needs
- Identifies awareness opportunities for promoting equal opportunities and other cross-curricular skills and themes
- Promotes topic planning across a range of curriculum areas
- Identifies assessment opportunities and methods of assessment
- Provides a permanent record of the class activities and any necessary follow-up
- Shows whether the week's activities cover a range of curriculum areas.

The weekly planning and assessment sheet was developed with the idea that it could actually replace a teacher's weekly record book, and is best used with all subjects to view (see page 34).

The weekly planning and assessment sheet is a very useful tool as it provides, at a glance, the work to be covered by a particular class or group. Again schools may wish to adapt it to meet their specific needs.

1 The lesson column is a reminder to the teacher of what is to be covered during that particular week, by whatever means are seen as appropriate. In some schools this could mean a title and brief explanation of each subject lesson with reminders of where detailed notes and worksheets can be located. In other schools the column could contain cross-curricular themes which cross subject boundaries for a more integrated approach.

2 Whatever the approach taken to completing the lesson column, it is suggested that the objective column is completed subject-specifically. Most objectives will be

Weekly planning and assessment sheet

22 .... to 26 January 1996

Subject	Lesson	Objective	Assessment	Future planning
English	Writing	To be able to write a report on the school sports day for inclusion in the newsletter. Accurate use of punctuation	P	
	Weekly spellings	To be able to spell this group's words	Q (AT3 rest of exic group)	
	Comprehension	To be able to understand 'Cat' poem	P	
Maths	Maths scheme	To be able to work independently, at own level, developing mental skills	P	
	Body maths (see notes)	To be able to measure pulse and breathing rates at rest and during exercise and record appropriately	O (notes to be made)	
Science	My body	To know position of heart, lungs, stomach, intestines, liver and kidneys	Q (written AT2 test at end of unit)	
		To understand how the heart pumps blood around the body	Supervisory marking	

Weekly planning and assessment sheet

22 .... to 26 January 1996

Subject	Lesson	Objective	Assessment	Future planning
History	Writing and Printing	To continue to develop a chronological understanding	Q (Level 3/4/5 test at end of unit)	
Geography	Map making	To be able to map accurately the area around school	O / P	
Art	Watercolour painting	To develop fine brush control	P	
Music	Singing and percussion 'Dem bones'	To be able to maintain a part independently of others	O	
RE	Rites of passage	To know about the rites of passage in Judaism	Q (test at end of unit)	
Technology	Body joints	To develop understanding of skeleton by constructing a jointed puppet	P	
IT	Redrafting	On-going use of computers to redraft own poem (see rota)	P	
PE	Games	Soccer - to be able to play as part of a team	O (notes to be made)	
	Gym (see notes)	To be able to sustain energetic activity	O (notes)	
Notes	Correct maths data in gym lesson. Get 'body' model for Wed.			

derived from the school schemes of work/ programmes of study and assessment. Recording and reporting are made much easier if subject boundaries are maintained.

(There is a specific requirement for the National Curriculum to be reported to parents in such a way that individual subjects can be identified.)

3 The assessment column can use the shorthand from the half-termly topic planning and assessment sheet (O, P, Q, S – see page 20). This can be supplemented with specific references to, for example, test question papers and groups or individuals requiring specific assessments.

4 The future planning column is left blank during the planning stage. It is suggested that this is completed during the week in question, as specific activities and assessments are undertaken. At its simplest, the column could record successful completion of lessons or units of work. Used in a formative way, the column can be an ever-accumulating record of individuals or groups who achieve, or probably more simply, do not achieve particular objectives. This information can then be fed into planning the next week's/half-term's work to ensure successful learning takes place. Seen as a collection of sheets for a half-term, they form the beginnings of a summative record for the class and for individuals.

5

EVALUATION

Earlier in this book we looked at ways of monitoring work taking place in the classroom. The areas we covered included appraising and recording the attainment targets visited by classes, and evaluating the progress of individual children to provide evidence for assessment throughout and at the end of key stages 1 and 2.

The science scheme itself also needs evaluating. In order to assess the effectiveness of our programme of learning we must build in review procedures. These should be considered at three levels:

- Individual activities
- Half termly topics
- The scheme of work.

Evaluating activities

A number of criteria need to be considered when selecting appropriate activities for the children.

Does the activity provide opportunity for the children to:

- Develop their understanding of the processes of science by experiencing skills?
- Foster the development of attitudes such as respect for evidence, perseverance, open-mindedness, cooperation, respect

for living things and self-criticism?
- Develop scientific concepts (Sc 2–4)?
- Reach a satisfactory outcome?
- Apply scientific ideas to real-life problems, especially in a technological context?
- Work cooperatively with other children?
- Communicate their ideas to others?

Does the activity:

- Stimulate curiosity?
- Relate to the interests and previous experiences of the children?
- Appeal to both boys and girls?
- Reflect multicultural aspects of society?
- Help the children to develop an understanding of the world around them through their interaction with materials?
- Involve the selection by the children, and their use, of a range of simple and safe equipment and materials?
- Use resources which are readily available in the classroom?
- Contribute to a broad-based, balanced science curriculum and build upon children's previous experiences?

By regularly assessing chosen activities to the above criteria you can ensure that you are meeting the children's entitlement to the science curriculum at the level of particular learning experiences. The use of topic or weekly planning and assessment sheets to identify the relevant skills and concepts and to evaluate the success of the activity within the classroom will provide all the necessary information.

However, no single activity can provide children with the full range of skills or the body of knowledge outlined by the National Curriculum, so the programme of learning needs to be evaluated at topic level.

Evaluating topics

When evaluating the topics you need to return to the criteria used in choosing the topic. These included:

- The needs and interests of the children, the appropriateness of the topic to their experience and stage of development and the time of year. (Were activities added to the topic because of the children's interests and enthusiasms? Did they enjoy the topic?)
- The content of the topic. (Is there a balance of coverage of the main areas of science?)
- The intended outcomes. (What did you expect the children to learn? Did the work actually cover the identified parts of the programmes of study? Did the children actually experience the skills expected?)
- Local resources. (Were the visits useful? Did you make the best use of parents, local groups, etc? Was the visit relevant to the needs of the children?)

The use of the sheets outlined earlier in the book will help to assess the topic as part of the learning programme. By transferring the information to the class record sheet you can better assess the success of the topic. This will help you to plan future work effectively for the children. You may find that you have to make some slight amendments to the next topic because of

over/under-coverage of programmes of study. This review of topics should be undertaken termly or half termly depending on the model chosen. The use of different coloured pens when completing the class record sheet will be useful when you reach the evaluation of the scheme itself.

Evaluating the scheme of work

The scheme of work should be seen as a working document. It reflects the school's thinking about the delivery of the curriculum, but it is not written in stone.

The sections within your scheme (outlined in Section B – for example, classroom organisation, equal opportunities, assessment) will need to be reviewed in the light of current legislation produced by the DfE, HMI, the NCC, SEAC, the LEA and your own school policy statements to take into account more appropriate and effective teaching methods, and also contemporary events which will bring more relevance to the children's work.

The chosen model should be evaluated separately. Again, you should return to the criteria used when the particular model was chosen. Each of the models outlined earlier had advantages and disadvantages and, no doubt through staff discussion and by working through your chosen model, you will have raised other core issues. By returning to those issues it will be possible to evaluate the model by undertaking an audit. Were the advantages reflected in actual practice? Have you found ways to remove the apparent disadvantages?

The timing of this review will again depend on the frequency at which you decided to cover the programmes of study. A realistic approach would be to review at the end of one year and then make the necessary amendments. To do so earlier would not allow enough time for reflection – to wait for too long, however, can lead to continuity problems.

You will need to evaluate the model again at the end of the two-year cycle using the same criteria.

Conclusion

The process of the development, implementation and evaluation of a scheme of work should be both gradual and cyclic. It is neither desirable nor realistic to attempt to reach a final product too rapidly. There must be full consultation with all staff throughout the stages of development. The aim should be to ensure the setting up of a process to evaluate and modify the scheme in the light of experience.

If this is done it will be possible to accomplish the aim set out in the following paragraph, taken from the final report of the *National Curriculum Science Working Group* (DES, August 1988):

> *"The planning and implementation of a balanced but flexible programme for Science which meets the requirements of the National Curriculum is a challenging but achievable goal."*

In many ways the development of a scheme of work is a journey on which it is as important to travel as to arrive. This book aims to provide helpful signposts on that journey!

B

TOPICS FOR CLASSROOM USE

This section provides a substantial bank of topic-based science ideas for classroom use. Each topic provides:

- An ideas bank drawn up by brain-storming ideas and reading the relevant part of the programme of study
- Linked activities
- Equipment and resources needed
- Extension ideas.

The topics outlined are based on the 'fixed and open' model. This model has been chosen as it is the one which appears most popular to primary teachers, the 'free' half term being used by many teachers to slot in specific history or geography based topics. The topics outlined have, however, been used by schools working on other models and can be adapted to fit any of the approaches outlined in Chapter 4. The topics will provide either an 'off-the-peg' scheme of work or flexible resources to meet individual needs.

The outlined activities, approximately ten to twelve per topic for key stage 1, and six to eight for key stage 2, are all science based and use only basic science equipment. The topics chosen are all broad-based and will thus facilitate a cross-curricular approach to the delivery of the curriculum.

This section has been divided into three parts:

- Infant topics, for classes R, Y1 and Y2
- Lower junior topics, for classes Y3 and Y4
- Upper junior topics, for classes Y5 and Y6.

In each section you will find references to the programmes of study that may be covered by the activities. This is not intended to be a comprehensive analysis. The exact knowledge, understanding and skills required from an activity are left for the teacher to judge.

- F = Foundation Science
- Sc = Science; followed by 1, 2, 3 or 4 and the appropriate section number and letter. For example, Sc2–3a refers to the statement: Pupils should be taught that plant growth is affected by the availability of light and water, and by temperature.

Note Key stage 1 and key stage 2 statements with the same reference are different!

6

KEY STAGE 1: INFANT TOPICS (RECEPTION, YEARS 1 AND 2)

Water

Keeping clean

- Sorting a collection of washing/cleaning materials
- Investigating ways of making bubbles
- Investigating which sponge holds the most water
- Investigating the best way to wash hands
- Designing and making a bathtime toy
- Investigating which soap or detergent is best for removing stains

Making water move

- Free play in the water tray
- Exploring different ways of making water move by pouring, squirting, using a wide variety of containers
- Investigating how far water can be moved using squeezy bottles
- Moving objects by squirting water on them (e.g. have a ping pong ball race; fair test idea: how far with 1, 2 or 3 squirts)
- Investigating siphons to make water move
- Investigating simple fountains
- Investigating designing, making and testing water clocks

Boats

- Exploring toy boats through free play in the water tray
- Designing, making and testing boats using Duplo®, Bauplay®, Plasticine®
- Investigating boat shapes, sails. Which travels furthest, fastest?
- Investigating boats that can carry cargo. How is the weight carried affected by the boat's shape, size?
- Designing, making and powering boats

Water play

- Investigating objects which float and sink
- Making floaters sink/sinkers float
- Investigating how to make milk bottle tops, Plasticine®, etc. float by changing the shape
- Investigating making water change shape
- Observing ice-balloons

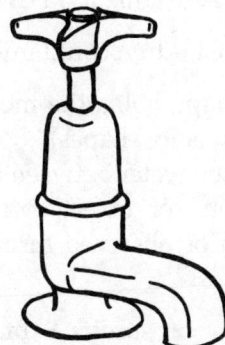

Water play activities

1 Floating and sinking Collect and test an assortment of objects which sink and float. Put them in sets.

Equipment: plastic tank, sponges, toys, cork, nails, wood, paper, Plasticine®, etc.

Extensions: children could investigate:

- Predicting before testing
- Whether or not all wood floats
- Making floaters sink.

Programmes of study: F–1a, 3a; Sc1–2b; Sc3–1a, 1b; Sc4–2b, 2c, 2d

2 Making floaters sink, sinkers float Can the children make Plasticine® or milk bottle tops float?

Equipment: plastic tank, Plasticine®, milk bottle tops, pieces of kitchen foil, etc.

Extensions: children could investigate:

- Changing the shape of their boat
- Adding cargo to try to make the floater sink.

Programmes of study: F–1a, 3a; Sc1–2b; Sc4–2b

3 Making water change shape Provide a variety of different shaped containers and ask the children to investigate ways of making water change shape.

Equipment: cylinders, plastic bottles, jugs, cups, etc. Plastic tank, food colouring.

Extensions: Children could investigate:

- Which shape holds the most water (use plastic capacity shapes)
- Sharing the water between containers
- The colour of objects when they are placed in or observed through coloured water.

Programmes of study: F–1a, 3a; Sc1–2b; Sc4–3a

Keeping clean activities

1 Sorting washing/cleaning materials Set up a collection of washing/cleaning materials and ask the children to sort them in as many ways as possible.

Equipment: brushes (e.g. nail, tooth, scrubbing), soap containers (e.g. soap powder, detergent, washing-up liquid, toilet soap), sponges, flannels, pan scrubs, loofah, etc.

Extensions: children could investigate:

- Matching object to purpose
- Materials objects are made of
- The amount of water different sponges or cloths hold.

Programmes of study: F–1a, 3a; Sc1–2b; Sc3–1e

2 Making bubbles What happens to different soaps and detergents when they are shaken with water in clear plastic bottles? How far up do the bubbles go?

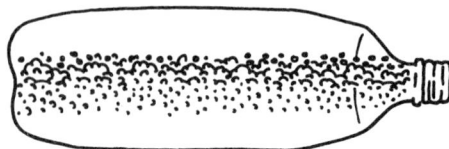

Equipment: soaps, shampoos, detergents, plastic bottles, water, stop-watch, florists' wire (to make bubble blowers), bubble blowers.

Extensions: children could investigate:

- Timing how long the bubbles last
- Fair testing by using the same volume of water, detergent
- Trying to make different shaped bubbles
- Observing what happens to the bubbles
- Making observational drawings of the bubbles
- The 'best' bubble solution
- Different ways of making bubbles.

Programmes of study: F–1a, 2a, 3a; Sc1–1c, 2b

3 Washing hands Find the best way to get hands cleanest: cold/warm water, with/without soap.

Equipment: plastic tank, cold water, warm water, variety of soaps (e.g. toilet, liquid, washing-up liquid), nailbrush, flannel, sponge.

Extensions: children could investigate:

- Which soap/detergent is best for removing stains such as tomato ketchup
- Fair testing using the same volume of water, soap powder.

Programmes of study: F–1a, 3a; Sc1–1c, 2b

Making water move activities

1 Moving water using different containers Provide a variety of containers. Encourage the children to find as many different ways as they can of making water move.

Equipment: water toys, pumps, bottles, tubing, funnels, sieves, straws, sponges, syringes, squeezy bottles, plastic tank/water tray.

Extensions: children could investigate:

- Using the water pump to fill different containers
- How far they can make water move
- Having a ping pong ball race in which they see who can move the ball farthest by squirting water on it.

Programmes of study: F–1a, 3a; Sc1–2b; Sc4–2c

2 Investigating syphons Provide plastic tubing with a variety of diameters (each about 50 cm long). Using a funnel, the children pour water into the tubing, holding an end in each hand, and they observe what happens to the water whilst raising and lowering alternate hands.

Equipment: plastic tubing, funnels, plastic tanks, boxes (to form an obstacle course), plastic sheeting.

Extensions: children could investigate:

- Using a syphon to transfer water from one container to another
- Making an obstacle course and using a syphon to make some water move over the course
- Fountains using straws, pen cases, etc.

Programmes of study: F–1a, 3a; Sc1–2b

3 Investigating water clocks After exploratory play with water timers, the children can design a simple water clock using a plastic squeezy bottle.

Equipment: plastic squeezy bottle, water, plastic tank, clamp or device to hold bottle steady, Plasticine®.

Extensions: children could:

- Investigate the number of times they can clap their hands, jump up and down, write their name, and so on, before the water clock empties
- Make surveys of the above activities when done by other class members
- Design and make a water timer.

Programmes of study: F–1a, 2a, 3a; Sc1–1c, 2b

Boat activities

1 Exploring toy boats Provide a collection of toy boats (clockwork, elastic driven). Ask the children to sort them and play with them in the water tray.

Equipment: toy boats, Duplo®, Bauplay®, Plasticine®, water tray, timers.

Extensions: children could investigate:

- Which boat travels farthest, fastest, and so on
- Designing and making boats.

Programmes of study: F–1a, 3a; Sc1–2b; Sc4–2c

2 Investigating boat shapes and sails The children investigate the 'best' boat by exploring the shape of the hull, different shaped sails.

Equipment: different boat shapes with dowel masts (e.g. round, rectangular, triangular), different shaped sails made of card, balloons, water tray, battery-powered fan, bellows, plastic guttering.

Extensions: children could investigate:

- How far the boat travels
- Which boat is fastest
- Fair testing by changing only one variable at a time
- Different ways of powering a boat
- Designing and making a paddle-boat.

Programmes of study: F–1a, 2a, 3a; Sc1–1c, 2b; Sc4–2a

3 Boats which carry cargo Design and make a raft which will carry a cargo. How much cargo will it carry ?

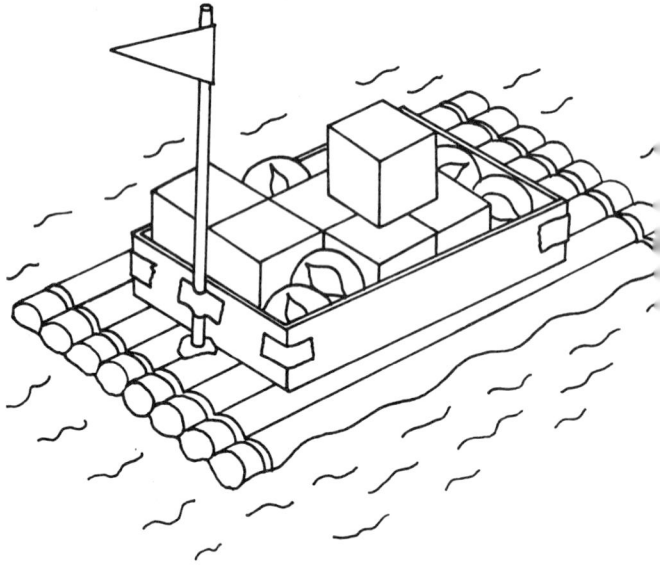

Equipment: squared paper, foil, Plasticine®, sticky tape, centicubes, dowelling, marbles, water tank.

Extensions: children could investigate:

- How much cargo the raft will carry before it capsizes
- The best place to put the cargo
- The most suitably shaped raft: broad-based and shallow, or narrow-based and tall-sided, etc.

Programmes of study: F–1a, 2a, 3a; Sc1–1c, 2b

My body

- How do I move? What can I do?
- Investigating hands and feet, making handprints, foot-prints
- Protecting our bodies
- Mixing paint to make skin colour
- Investigating fingerprints

I am growing

- How am I different from when I was small?
- What can I do now that I couldn't do then?
- Which clothes will fit me?
- What are my favourite foods?
- Which are good for me?
- What do I eat for breakfast?
- Does everyone eat the same?
- Investigate different fruits and vegetables. How do they grow?
- How important is keeping ourselves clean?

Me and others

I can...

- See to sort objects by colour, size and shape
- Investigate objects which help us to see better
- Look for clues
- Play Kim's game
- Make noises
- Muffle sound
- Make string telephones
- Sort objects by sound
- Send messages by touch
- Feel with my feet
- Taste different foods
- Smell different foods

Playing safely

- Setting up a clinic/hospital
- Investigating safe toys
- Looking at playgrounds, visiting parks
- Looking at surfaces. Are they safe in wet/dry weather?
- Making a model swing
- Designing an adventure playground
- Investigating shoes. Which have best grip?
- Balancing
- Bouncing balls

My body activities

1 How do I move, what can I do? Look in a mirror. How many ways can you move different parts of your body? Start with your head (eyes, eyebrows, tongue, mouth, nose, ears). Then move lower down and see which other parts of your body you can move.

Equipment: mirror, hand lens, paints, crayons, paper.

Extensions: children could investigate:

- Working with a partner to make sad, happy, cross faces. They could try to copy their partner's expressions and draw or paint sad/happy faces
- Sorting the types/colours of hair in the class
- Mixing paints to get as near a match as possible to their skin colour
- Naming as many parts of the body as

possible, drawing round a child and then labelling parts, joints, and so on.

Programmes of study: F–1a, 3a; Sc1–2b, 2c; Sc2–2a, 4b

2 Investigating hands and feet Working with large sheets of painted paper or in a sand pit, children make footprints – standing, walking, running. How are they the same? How different?

Equipment: paints, paper, sand pit, hand lens, access to a sink, soaps, different textures (e.g. sandpaper, cottonwool, fur, wool, stone).

Extensions: children could investigate:

- Looking closely at their hands to see where dirt collects
- Finding the best way to clean hands (e.g. using hot/cold water, different soaps) then looking at hands again and seeing how clean they are
- Touching skin with different objects. Trying different areas of skin (e.g. palm, back of hand, knee, cheek, fingernail) and describing what it feels like. This can be turned into a game by working with a partner who is wearing a blindfold (Can you guess which object is being used?)
- Making fingerprints and comparing with others. How are they the same? How different?

- Ways to protect skin (e.g. using gloves, creams)
- Making a machine which will pick up objects.

Programmes of study: F–1a, 2b, 3a; Sc1–2b; Sc2–4b

3 Making skin colour Provide a variety of colours of paints and ask the children to make their skin colour.

Equipment: paints, mirror, hand lens, PVA glue to thicken the paint.

Extensions: children could investigate:

- Different thicknesses of paint
- Making different shades of the same colour
- Hot and cold colours
- Other similarities and differences – i.e. hair colour, eye colour
- Making self portraits.

Programmes of study: F–1a, 3a; Sc1–2b; Sc2–4b

I am growing activities

1 How am I different now? The children bring in baby photographs to compare with themselves now. Discuss what they can do now compared with what they could do as babies.

Equipment: mirror, photographs, paper, paints, scissors, clothes from the dressing up box, baby clothes, etc.

Extensions: children could investigate:

- What they can do with their feet
- Making outlines of feet and shoes; do shoes fit well?
- The clothes in the dressing up box, comparing them with the baby clothes. (Which clothes could they wear, trying them on; finding out how they know which clothes will fit them – by looking, trying them against themselves, looking at the size labels, and so on)
- If the tallest person has the biggest feet
- If the person with the largest hands can pick up the most objects.

Programmes of study: F–1a, 3a; Sc1–2b; Sc2–2a, 4b

2 The food we eat Make a collection of breakfast foods including rice, fruit, etc. Taste them and look at the ingredients. How do we eat them? What happens when we put milk on cereals?

Equipment: breakfast cereals, fruit, pictures of food from magazines, Plasticine®, clay, Play-doh®, dough, bread, access to cooker, different breads, toothbrushes, toothpaste, disclosing tablets.

Extensions: Children could investigate:

- Food in different forms (e.g. fresh, frozen, tinned)
- Making toast, cooking for different lengths of time, using different breads
- Comparing breakfasts from around the world
- Cutting out pictures of favourite foods, making surveys
- Sorting pictures of foods which are good for you
- Making models of their favourite meals
- Sweets and their teeth. Use a disclosing tablet after eating a sweet. Draw what happens. Clean teeth thoroughly, then draw another picture.

Programmes of study: F–1a, 2b, 3a; Sc1–2a, 2b; Sc2–1b, 2a, 4b

3 Fruit and vegetables Provide a collection of fruit and vegetables (including exotic varieties). Ask the children to talk about the fruits, sort them, make observational drawings, taste the fruit.

Equipment: fruits of various colours, shapes and textures, hand lens, binocular microscope, charcoal, paper, crayons, etc.

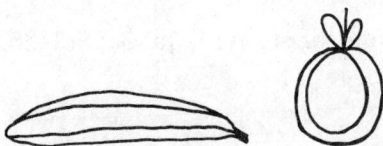

Extensions: children could investigate:

- Which fruits roll
- Which fruits sink/float
- Different tastes by conducting a survey to find favourite fruits
- The seeds of the fruits.

Programmes of study: F–1a, 3a; Sc1–2a, 2b, 3a; Sc2–3b, 4b, 5a

I can... activities

1 Objects which help us to see better Explore a collection of aids which help us to see better. Match objects to their purpose.

Equipment: spectacles, telescope, hand lenses, mirrors, binocular microscope, torches, periscope, acetate sheets.

Extensions. Children could investigate:

- Making observational drawings of a range of objects after using hand lens or microscope
- Painting self-portraits using mirrors
- Symmetry
- Making coloured spectacles, observing different coloured objects
- Setting up an optician's in the structured play area, designing eye charts, etc.

Programmes of study: F–1a, 3a; Sc1–2b, 2c; Sc2–4b

2 Making noises Investigate how many different ways sounds can be made: by plucking, tapping, blowing, scraping, etc.

Equipment: musical instruments, yoghurt pots, pulses, elastic bands, rulers, containers, greaseproof paper, art straws, etc.

Extensions: children could investigate:

- Making their own musical instruments
- Playing a sound game using sealed boxes and a tester box
- Muffling sounds, making ear-muffs.

Programmes of study: F–1a, 3a; Sc1–2b; Sc4–3c, 3e

3 Sorting objects by touch Make a touch collection. Ask the children to describe what each object feels like. Then ask them to sort the objects just by touch.

Equipment: variety of objects which can be distinguished by touch, 'feely' bags, charcoal, crayons, paper.

Extensions: children could investigate:

- Sorting a collection of different kinds of paper by touch
- Different water temperatures
- Going on a touch walk, making rubbings of different surfaces
- Trying to identify objects by touch when wearing gloves
- Designing and making a message or sign which could be read by a blind person.

Programmes of study: F–1a, 3a; Sc1–2b

Playing safely activities

1 Safe toys Visit the local baby clinic. Set up a clinic/hospital/dental surgery in the play area. The children bring in a collection of toys. They choose toys suitable for taking to bed, for playing with outside, for a younger child, for playing with in the bath, and so on.

Equipment: selection of toys, junk material, fabrics, glues, construction kits.

Extensions: children could investigate:

- Making toys for younger children, rolling toys, children who cannot see, children who cannot hear, etc.

Programmes of study: F–1a, 3a; Sc1–2b

2 The playground Go to the park or playground with the children and look at the surfaces. Feel them, stamp on them. Try bouncing a ball on them. How do children think that different weather conditions will affect them (ice, wet leaves, rain, snow)?

Equipment: balls, Duplo®, Lego®, junk materials, glue, Plasticine®.

Extensions: children could investigate:

- Making a model swing using Duplo®, Lego®, adding safety or double seats
- Designing an adventure playground using junk material. (Conduct surveys to discover if other children agree with their choice of equipment)
- The safety aspects of playgrounds – surfaces, angle of slope for slides.

Programmes of study: F–1a, 2a, 3a, 5a; Sc1–1c, 2b, 3b

3 Investigating shoes Which shoes are the best for running? Take sole rubbings or prints. Which shoes have the best grip?

Equipment: shoes with a variety of soles, paper, paints, slope with different surfaces (e.g. polished/unpolished, metal, carpet).

Extensions: children could investigate:

- Which surface is best for a slide
- The best angle of a slope for a slide
- If the slide works as well when wet.

Programmes of study: F–1a, 3a; Sc1–2b, 3c; Sc3–1e

Soft toys

- Compare and sort by colour, size, weight
- Make teddy's hat
- Make a teddy bears' picnic

Dolls

- Dressing dolls for summer/winter, different cultures
- Washing dolls
- Naming parts of the body
- Making a hat for a giant

Toys made with magnets

- Investigate magnets
- Make pictures using iron filings
- Design and make paper clip sculptures
- Design and make magnetic games

Puppets

- Investigate different kinds of puppets
- Make a jointed puppet
- Make a puppet theatre
- Shadow puppets

Flying toys

- Design and make autogyros, planes
- Can you make your flying toy better by changing size, materials, weight?

Mechanical toys

- Collect and sort
- Investigate what gives them their 'go'
- Wind up toys, springy toys

Toys which make sounds

- Collecting and sorting
- Design and make musical instruments which can be blown, shaken, plucked
- How far away can we hear sounds?
- Design and make a sound box (a shoe box guitar)

Construction toys

- Lego®, Duplo®, Capsela®, Georello®

Activities with toys

1 Soft toys, dolls and puppets Look at a collection of toys: soft, dolls, puppets and choose your favourite. Match toys to their purpose. Compare and sort by size, type, material.

Equipment: toys, fabrics, dowel, paper bags, glues, scissors, pan balance.

Extensions: children could investigate:

- Placing the toys in feely bags and identify by touch alone, or describing what they can feel to a partner who then has to identify the toy
- Making a variety of puppets using paper bags, glove puppets, finger puppets, stick puppets, jointed puppets, and so on depending on the age and experience of the children. They could use the puppets to make up plays

- Setting up a clinic in the play area after a visit to the local clinic
- If the biggest toys are also always the heaviest, weighing, sequencing, and so on
- Materials for toys, matching material to purpose, etc.

Programmes of study: F–1a, 3a; Sc1–2a, 2b; Sc3–1e

2 Dressing toys Ask the children to dress the toys for different occasions or seasons and explain their choices.

Equipment: fabrics, waterproof materials, glue, scissors, foods, paper, toy trucks.

Extensions: children could investigate:

- Making a sun hat for teddy, making a rain hat for teddy, identifying similarities/differences
- Finding out which of the toys will go down a slope the fastest
- Making a hat for a giant.

Programmes of study: F–1a, 3a; Sc1–1c, 2b, 3d; Sc4–2c

3 Making a teddy bears' picnic Plan and prepare a teddy bears' picnic and discuss the healthy eating aspect.

Equipment: pictures of food from magazines, Play-doh®, clay, paints, paper plates, food items as appropriate.

Extensions. children could investigate:

- Their favourite foods, by conducting surveys, for example
- Preparing a picnic.

Programmes of study: F–1a, 4a; Sc1–2b, 2c, 3b

4 Puppets Look at different kinds of puppets – finger, glove, paper bag puppets. The children then make their own puppets.

Equipment: felt, paper bags, paper plates, gloves, fabric, tights, dowel, glue, string.

Extensions: children could investigate:

- Making a jointed puppet
- Comparing toy and human joints
- Making shadow puppets, writing plays.

Programmes of study: F–1a, 3a; Sc1–2b, 3c; Sc2–2a

5 Toys which make sounds Make a collection of toys which make sounds. Sort the toys according to how the sound is made, i.e. by blowing, hitting, winding up.

Equipment: toys, junk materials (e.g. wood, plastic, metals), tape recorder.

Extensions: children could investigate:

- Matching recorded sounds to toys
- Sorting sounds into loudest/softest, highest/lowest
- Making a 'sound' collection, using objects hung on clothes-line across the classroom, using different materials.

Programmes of study: F–1a, 3a; Sc1–2a, 2b, 3c; Sc4–3c, 3d, 3e

6 Making sounds Investigate ways of making sounds – plucking, shaking, scraping, hitting, blowing.

Equipment: various musical instruments, shakers, dried peas and beans, rice, cotton reels, sand, beads, fabrics, elastic bands, greaseproof paper, sticky tape, ruler, junk, split pins, etc.

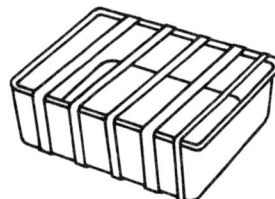

How to make a Guitar

Extensions: children could investigate:

- Music from elastic bands by plucking them, trying several widths and lengths.
- Making drums using different shaped containers (e.g. plastic, wood, metal), investigating different 'skins'
- Making shakers using a variety of contents. This could be made into a matching game where the children work in twos and one child has to identify the contents of the shaker
- Playing a tune which other children can recognise.

Programmes of study: F–1a, 3a; Sc1–2a, 2b, 3c

7 Making sound boxes Design and make a shoe-box 'guitar'. Does the size of the hole make a difference to the sound?

Equipment: shoe boxes, variety of elastic bands, split pins, balsa wood.

Extensions: children could investigate:

- If adding a 'fret' makes a difference to the sound
- If the thickness/length of the elastic band makes a difference.

Programmes of study: F–1a, 3a; Sc1–3c, 3e, 3f; Sc4–3c

8 Magnets Provide a play table for magnets. Explore a variety of different shaped magnets. Provide a series of magnetic games – catching fish, placing the donkey's tail, and so on.

Equipment: a selection of different shaped magnets, paper clips, selection of materials (magnetic/non-magnetic), scissors, glue.

Extensions: children could investigate:

- Sorting materials into those which are/are not attracted to a magnet
- The number of paper clips a magnet will hold

- Making a paper clip sculpture using magnets
- Making a paper clip walk across a variety of surfaces (e.g. a piece of paper, a sheet of perspex, balsa wood)
- What happens when you try to put two magnets together?

Programmes of study: F–1a, 3a; Sc1–3c; Sc3–1b

9 The strength of magnets Ask the children to find out where a magnet is stronger – in the middle or at the ends?

Equipment: magnets, paper clips, iron filings, Petri dishes, card, metal discs.

Extensions: children could investigate:

- Which metals are/are not attracted to a magnet
- Making identikit pictures with iron filings, adding hair, beard, moustache to faces. (It is safer to do this using Petri dishes sealed with sticky tape. The children draw the faces, then you add the iron filings and seal the Petri dish.)
- Designing and making board games using magnets.

Programmes of study: F–1a, 3a; Sc1–2a, 2b, 3c; Sc3–1b

10 Mechanical toys Collect and sort a collection of mechanical toys. Try to find what gives them their 'go'.

Equipment: selection of toys (e.g. clockwork-operated, battery-operated, motor-operated toys, jack-in-the-box, spinning tops), timers.

Extensions: children could investigate:
- How 'efficient' the toys are
- Designing and maing 'jumping frogs' and other similar toys.

Programmes of study: F–1a, 1b; Sc1–2b, 3c

51

11 Construction toys Ask the children to make a toy for a younger child.

Equipment: construction kits, e.g. Meccano®, Lego®, Duplo®, Capsela®, Georello®, etc.

Extensions: children could investigate:

- Ways in which they can make the toy move
- Conducting a survey to find favourite toys.

Programmes of study: F–1a, 3a; Sc1–2a, 2b; Sc4–2d

Look up, look down, look around

Buildings

- Go for a buildings walk
- Different homes for different animals, insects, etc.
- Homes around the world
- Make a Lego® building site
- Which mixture makes the best cement?
- Construct a tall, strong tower
- Tessellations. Which shapes fit best together?

Seasons

- Make seasonal collections
- Celebrations around the world
- How the animals adapt to the weather, e.g. home, coat, hibernation, migration
- Seasonal foods, e.g. fruit, seeds, vegetables
- Clothes – differences in materials, e.g. compare thicknesses
- Does one wooly jumper keep you as warm as lots of layers?

Light and dark

- 'Light spotting' walk
- Sorting shiny/dull, light/dark, bright/dim objects, materials
- Looking through coloured glasses
- Which colours can you see in the dark?
- Making and measuring shadows
- Using mirrors e.g. periscope, kaleidoscope
- Collecting things that reflect light

Look up,
Look down
Look around

In the park and garden

- Explore the school garden, local park
- Design and make a junk playground with movable apparatus
- Find the best surface to make a slide
- How many different ways can you move on the apparatus in the park?
- A minibeasts survey
- Make your own tray garden
- Planting seeds in different conditions

The weather

- How does the weather affect me?
- Clothes for keeping cool, warm, dry
- Activities I can do in the rain, in the sun
- Food I eat when I am hot, cold
- Weather records
- Measuring temperature
- Keeping things warm/cold
- Washing day
- Measuring the weather, e.g. rain gauges, wind direction
- Keeping dry: rain drops, puddles, waterproofing
- Ice: how long can I keep an ice cube?

Time

- How long does it take to... tie shoe laces, run round the playground, write name?
- Devise a time line of your day
- Diary of activities – school day, weekend
- Collect and sort timing devices
- Survey on bed times, breakfast times

Buildings activities

1 A building walk Take the children on a walk to look at the buildings around school and the neighbourhood. Compare new/old, uses of buildings, shapes, patterns, textures, colours. Visit a building site by arrangement. What materials are used? What machines can you see?

Equipment: bricks, cement, sand, plaster, wood, card, straw, bulbs, batteries, wires, Duplo®, Lego®, building blocks, photographs of buildings, Humpty Dumpty.

Extensions: children could investigate:

- Story book houses to find the best, e.g. The Three Little Pigs, Hansel and Gretel, The Mice who Live in a Shoe. They could design and make the houses and think of ways in which to test them
- Building houses for their toys, adding doors, windows, lights, and so on
- Sorting building materials by shape, colour, texture, usage
- Testing various building materials, e.g. straw, wood, bricks for strength in wind, rain and against pushes and pulls
- Setting up a play building site in the play area, making their own bricks, designing and making machines
- Making a den for themselves and their friends
- Building a wall strong enough to seat Humpty Dumpty using building blocks, Duplo®, etc. They could devise ways of testing its strength
- Drawing plans of their classroom, looking at building plans, drawing plans of a room at home, looking at similarities/differences
- Using paper straws/pipe cleaners to make structures, investigating strong shapes
- Ways of moving a brick using slopes, construction toys, wheels, rollers, etc.

Programmes of study: F–1a, 1b; Sc1–1c, 2b, 3f; Sc4–2a, 2b

2 A tall strong tower Construct a tall tower using the materials provided.

Equipment: 30 plastic straws, 1 m of sticky tape, 10 pipe cleaners, a metre ruler.

Extensions: children could investigate:

- The weight which the tower will hold.

Programmes of study: F–1a, 3a; Sc1–1b, 1c, 3c

Seasons activities

1. Seasonal collections of fruits, seeds, vegetables Ask children to make seasonal collections of natural and made objects. Look at the collection closely and draw, paint or model an object.

Equipment: leaves, flowers, fruits, seeds, clothes, items for celebrations, hand lens, microscope, paints.

Extensions: children could investigate:

- Matching objects/pictures to the seasons
- Designing/making celebratory objects
- Collecting seeds, fruits and leaves. Sort them in different ways, ordering, matching seed to fruit to leaf to plant using a clue book.

Programmes of study: F–1a, 3a; Sc1–2b, 3c; Sc2–4b

2 Season-spotting walks Take the children out for a walk round the neighbourhood. Ask them to search for clues which identify the season and record their results (drawings, photographs). Repeat the walk one season later and compare the results.

Equipment: camera, paper.

Programmes of study: F–1a, 3a; Sc1–3c, 3d, 3e

In the park and garden activities

1 Exploring the garden Explore the school garden, grounds or local park. Ask children to collect fallen leaves, petals, seeds, and make a collage. Ask them to try to make a coloured picture without using paints, i.e. by rubbing with natural materials, such as petals and leaves, on to paper or cloth.

Equipment: leaves, petals, seeds, hand lens, microscope, different growing media, litter.

Extensions: children could investigate:

- Growing seeds in a variety of media, measuring growth rates and subjecting plants to different conditions (light/dark, with/without water)
- Designing a machine which will help to separate seeds by size
- Planting sunflower seeds and comparing growth to their own height
- What happens to different kinds of litter (e.g. garden refuse, plastic, cans, paper, fruit peel) if they are buried/are wet/dry over a period of time.

Programmes of study: F–1a, 1b, 3a, 4b; Sc1–1a, 1b, 1c, 2c, 3e; Sc2–3c

2 Mini-beasts Collect minibeasts. Ask the children to look closely at the creatures and make observational drawings, models.

Equipment: minibeasts, containers, pooters, brushes, hand lens, midi-spectors, mini-spectors, etc.

Extensions: children could investigate:

- A minibeast hunt – finding a minibeast (e.g. slug, worm, woodlouse) and then observing how it moves, making observational drawings and/or making models.
- Finding out where minibeasts like to live by setting up choice chambers – dark/light, damp/dry. The children could

think up a test which is fair and kind to their minibeast. (The children should always be encouraged to return their minibeasts to their natural habitat after investigations.)

Programmes of study: F–1b, 2a, 2c; Sc1–3f; Sc2–5b

Light and dark activities

1 Using mirrors Ask children to look at themselves in a plane/concave/convex mirror. They should draw what they see.

Equipment: plastic mirrors, cardboard tubes, washing up liquid bottles.

Extensions: children could investigate:

- Using mirrors to make a kaleidoscope, hinging the mirrors to see how many images they can obtain
- Practising mirror writing, devising codes, and so on
- Making periscopes.

Programmes of study: F–1a, 1e; Sc1–2b

2 Colour and light sources Look for sources of light on a 'light spotting' walk (e.g. street lights, car headlights, traffic lights, hazard lights). How are they the same? How different?

Make a collection of objects or pictures of objects which give out light (e.g. torch, candle, light bulb, sun).

Equipment: torches, bulbs, wires, switches, cellophane, spectacle frames, shiny/dull objects, opaque/transparent materials.

Extensions: children could investigate:

- Using a torch to sort objects as shiny/dull, opaque/transparent
- Making a transparent material opaque
- Making a torch, adding a switch
- Making a light in a peep-hole box and

using this to investigate which colours can be seen in the dark

- Looking through coloured glasses, using cellophane to make 'lenses'. They could then record what colours they see and see if coloured objects look different through coloured lenses.

Programmes of study: F–1a, 3a; Sc1–2b; Sc4–1c, 3a

Weather activities

1 Weather records Keep simple weather records daily and weekly. What sort of clothes do we wear at different times of the year?

Equipment: weather charts, thermometers, junk materials, water, pipettes, collection of fabrics.

Extensions: children could investigate (at suitable times of the year):

- Making raindrops using pipettes or straws, dropping them on to different fabrics and observing how the drops behave
- Making a container to catch the rain
- Making a device to measure the strength and direction of the wind
- Looking at a thermometer, seeing if they can make it go up/down, observing what happens when they take it outside
- Predicting then testing what happens when water is dropped on to different surfaces, e.g. flags, tarmac, grass, gravel, soil. They could draw round puddles, revisit them at intervals, make suggestions about what happens to the water.

Programmes of study: F–1a, 3a; Sc1–1b, 2c; 3e

2 Investigating an ice cube Ask the children to predict how long they can make an ice cube last. Ask them to suggest ways in which they can keep it frozen for a longer period.

Equipment: ice cubes, newspaper, straw, polythene, plastic, fabrics, timers.

Extensions: children could investigate:

- Fair testing by changing only one variable at a time
- Ice-balloons
- A container of snow – guess and then observe how long it will take for all the snow to melt.

Programmes of study: F–1b, 2a; Sc1–1c

Time activities

1 Time-lines Ask children to make a time-line of their day. Compare it with a friend's. Observe with care the position of the sun at intervals during the day. Compare shadow size and shape at different times during the day. Make and keep a record of whether it was dark/light when getting up/going to bed, whether the moon was visible last night, what shape it was.

Equipment: paper, drawing, writing materials, clock, large sheets of paper to draw round shadows.

Extensions: children could investigate:

- Playing with shadows, shadow puppet shows, making shadows with their hands, using a shadow box, silhouette pictures.

Programmes of study: F–1a; Sc1–3c, 3d; Sc4–3b

Celebrations

Growing up

- Identifying people from baby photographs
- Collecting facts about one another – height, weight, shoe size, birthdays
- Discussing activities which children can do now which they couldn't as babies
- Identifying growth patterns in people, animals
- Inviting elderly people into school to find out how they celebrated special events

Parties

- Dancing – investigating how different parts of the body move
- Music – designing and making musical instruments
- Food – making celebration sweets, observing changes
- Designing menus for special parties – Hallowe'en, birthdays
- Presents – investigating 'mystery' presents, matching presents and recipients

Decorations

- Sorting and matching decorations to celebrations
- Investigating which materials would be best for flags, bunting

Lights and candles

- Investigating mirrors
- Making shadow puppets
- Using fairy lights. Identifying colours, threading beads in the same order as lights
- Investigating candles

Clothes

- Identifying the sequence of events in getting dressed
- Looking at celebration clothes worn by different cultures
- Dressing up in clothes for different festive occasions

Growing up activities

1 Who is it? Make a collection of baby pictures of children and staff. Can children guess who they are?

Equipment: photographs, squared paper.

Extensions: children could investigate:

- Devising a chart which identifies when everyone has a birthday

- Devising questions for other children to find answers using the graph.

Programmes of study: F–4b; Sc1–3b

2 What can we find out about each other? The children can collect information about each other, e.g. hair colour, eye colour, height, weight, shoe size, and could make graphs and charts.

56

Equipment: squared paper, metre rules, bathroom scales.

Extensions: children could investigate:

- Relationships between height, weight, shoe size.

Programmes of study: F–3a; Sc1–2c; Sc2–1b

Parties activities

1 Musical instruments Design and make musical instruments and look at a collection of multicultural musical instruments. Can children identify by the sound alone which one is being played?

Equipment: junk materials to make musical instruments (e.g. elastic bands), multicultural instruments, tape recorder.

Extensions: children could investigate:

- Writing and playing tunes.

Programmes of study: Sc1–3c

2 Food Make celebration sweets using icing sugar, egg white, flavouring; or coconut, condensed milk, colouring, and so on. Observe what happens when ingredients are mixed together.

Equipment: various foodstuffs (e.g. chocolate, cornflakes), selection of exotic fruits and vegetables, recipe books, pictures of celebration meals.

Extensions: children could investigate:

- Tasting fruit, making observational drawings of the insides and outsides, conducting preference tests
- What happens to fruit when it is placed in water (e.g. bobbing apples)
- Making a collection of or pictures of foods for special occasions. (Remember to include celebrations of different cultures.) They could find ways of sorting the foods
- Decorating eggs, rolling them down slopes, and so on
- Devising recipes for parties, celebration meals, etc.

Programmes of study: F–1a; Sc1–2a

3 Presents Ask children to make a collection of boxes to hold a fragile present. Test them and compare them for strength.

Equipment: boxes, materials for packing, weights, parcels with 'mystery' contents, chalk sticks (for the fragile present), wrapping papers.

Extensions: children could investigate:

- The best container for sending a fragile present through the post
- Guessing mystery presents by shaking boxes or by feeling inside
- Ordering parcels by size, weight. Is the largest parcel always the heaviest?

Programmes of study: Sc1–3c

Decoration activities

1 Sorting and making Ask children to sort a set of material suitable for making decorations. What were their reasons for choosing them? Make a circuit to light up a lamp and use it to make a decoration. (It could be used as a Diwali lamp.)

Equipment: bulbs, batteries, wires, scraps of material, coloured papers (e.g. tissue, cellophone), scissors, glue, junk.

Extensions: children could investigate:

- Making decorations for specific occasions during the year
- Making kites for Songkran, etc.

Programmes of study: F–1a; Sc1–2b

2 Making cards The children design and make cards for special occasions. (Remember to include multicultural celebrations.)

Equipment: card, glue, wire, batteries, bulbs, springs, wood, material.

Extensions: children could investigate:

● Designing and making pop-up cards
● Cards with moving parts
● Cards which light up.

Programmes of study: F–1c; Sc1–3a; Sc4–1b

Activities with lights and candles

1 Patterns of colours Use a set of Christmas tree lights. Ask the children to identify the colours and thread beads in the same sequence as the lights.

Equipment: Christmas tree lights, threading beads, paper, tissue, paints, felt-tips.

Extensions: children could investigate:

● Making up their own colour sequences for other children to copy.

Programmes of study: Sc1–3c

2 Candles Make a collection of candles. Children compare them for length, thickness and weight. How long do they take to burn down?

Equipment: candles, timer, wire, bulbs, battery, tissue paper, cardboard cylinders.

Extensions: children could investigate:

● Predicting how long the candle will take to burn down before testing
● Making candle clocks
● Designing and making model candles using bulbs, etc.

Programmes of study: F–3a; Sc1–1b; Sc4–1b

Clothes activities

1 Special clothes for special occasions The children dress up in clothes for different festive occasions. How are they the same? How are they different? Could they go outside in them in hot/cold/windy/rainy weather?

Equipment: clothes from the dressing up box (remember to include clothes from a range of cultures), pictures of clothes, junk, paper, glues, scissors.

Extensions: children could investigate:

● Collecting and sorting, using their own criteria, pictures of clothes for special occasions
● Making fancy hats for a dolls' and teddies' parade, making very tall, wide hats, and so on
● Making carnival costumes from a wide variety of junk materials.

Programmes of study: F–1a; Sc1–3c

Wheels and toy cars

- Investigating wheeled objects and toys: how many wheels, where are they?
- Investigating the tread of tyres, first by observation, looking at the patterns, then by trying to add tread to wheels they have made.
- Designing, making and testing wheeled vehicles using criteria such as size, material shape
- Investigating toy cars and how they move: on the flat, on the slope, on different surfaces

Pushing, pulling, rolling and sliding

- Making pliable materials move by stretching, rolling, pulling
- Testing the elasticity of certain materials
- Investigating how different shapes move when they are rolled
- Investigating how surfaces affect the way objects move
- Designing and making a marble run, switchback
- Investigating moving large objects

Moving on land

Using machines

- Investigating simple cogs, making and testing home made cogs
- Investigating pulleys used in toys, e.g. building and testing cranes
- Investigating household tools, e.g. those which use cog wheels
- Investigating and making models of machines which help us to lift heavy objects.
- Using construction kits to design and make machines
- Designing and making imaginary machines
- Magnets as machines

Balancing

- Trying to balance different objects using a simple seesaw balance
- Investigating how different shapes balance
- Designing and making: balancing toys, pop-up toys, mobiles

Activities with wheels and toy cars

1 A collection of wheeled objects Ask the children to make and sort a collection in different ways. How are they the same? How different?

Equipment: collection of wheeled objects, metre rule, pictures of wheeled vehicles, paints.

Extensions: children could investigate:

- Sorting pictures of wheeled vehicles by the number of wheels, etc. and predicting which would be the easiest to drive
- Predicting and testing which wheeled vehicle will travel furthest with one push (how can they make the test fair?)
- Making tyre patterns by painting the tyres on toy cars and comparing them.

Programmes of study: F–1a, 3a; Sc1–1a, 1b, 1c; Sc4–2a, 2b

2 Making wheels Can children make a wheel? How does their wheel move? Is its path straight? What difference does the shape of the wheel make to its movement? What about the size of the wheel?

Equipment: Plasticine®, card, balsa, straws, junk, glue, scissors, card/plastic/wooden wheels (for extension work).

Extensions: children could investigate:

- Making a wheeled vehicle (by adding a simple axle and card, plastic or wooden wheels to a cardboard box), looking at how the vehicle moves and what affects its movement
- The best shape for a wheel by making wheels of different shapes and observing how they move
- Making a wheeled vehicle, adding tread to the wheels, changing the size of the wheels.

Programmes of study: F–3a; Sc1–3e

3 Toy cars The children investigate toy cars. Which one travels furthest on the flat? Is the result always the same? Is the test fair?

Equipment: toy cars, timer, metre rule, weights, rigid board and stand for slope, carpet tiles, polythene, sand paper, water (to change the surface).

Extensions: children could investigate:

- If adding a weight makes a difference to the distance travelled
- Trying different surfaces on which to travel
- How the car travels down a slope, predicting what might happen if the angle of slope is altered.

Programmes of study: F–1c; Sc1–1a, 1c; Sc4–2a, 2b

Pushing, pulling, rolling and sliding activities

1 Pushing and pulling The children investigate pliable materials, such as Play-doh®, Plasticine®, clay, Blu-Tack®, dough. Can they make a figure or a shape? Which is the best material? Are some materials too hard or too soft?

Equipment: materials as above, rolling-pin, cylinders, boards, shaped pastry cutters, ruler, centicubes.

Extensions: children could investigate:

- Predicting which materials are the stretchiest, then test their guess
- The differences between materials when stretched by hand and by being rolled out.

Programmes of study: F–1a; Sc1–3e, 3f; Sc4–2d

2 Moving heavy objects Investigate different ways of moving a large box along the floor. The children should try using rollers, pushing and pulling.

Equipment: large box, rope, rollers, brick, string, Newton meter, different surfaces, pencils, cotton reels.

Extensions: children could investigate:

- How many ways they can find to move the brick, trying to move it on different surfaces.

Programmes of study: F–1b, 1c; Sc1–2b; Sc4–2c

3 Rolling and sliding Provide a selection of three-dimensional shapes. Ask the children to roll these down a slope. Which shapes roll best? Sort the shapes.

Equipment: slope, 3D shapes, plastic containers, sand, water, Plasticine®, materials to change the surface.

Extensions: children could investigate:

- Predicting which objects will roll and which will slide, before testing
- Adding water, sand or Plasticine® to the containers and observing the effects
- Changing the surface of the slope.

Programmes of study: F–1a; Sc1–3c; Sc4–2a, 2b, 2c

Balancing activities

1 Balancing toys Make a collection of balancing toys and ask the children to sort them. Will they stand upside down?

Equipment: selection of balancing toys brought in by the children, cotton thread, wire coat hanger, length of garden cane, card shapes, ping pong balls, Plasticine®, matchsticks, glue, scissors.

Extensions: children could investigate:

- Making a simple mobile with two card shapes and the garden cane
- Making a more complex mobile and using Plasticine® to make it balance
- Making a balancing bounce-back toy, then trying to get it to bounce back if tapped gently but to stay down if tapped firmly.

Programmes of study: F–1a; Sc1–2a

2 Balancing objects Using various three-dimensional shapes, ask the children to investigate how they balance – on one edge, one face, etc. Ask them what happens when they turn the shapes around.

Equipment: 3D shapes, matchboxes, rulers, Lego® pieces, beads, marbles.

Extensions: children could investigate:

- Making seesaws, placing one object in one pan and using other objects to try to balance the seesaw
- Varying the position of a pivot on a seesaw which they have made.

Programmes of study: Sc1–2b; Sc4–2b

Using machines activities

1 Investigating cog wheels The children make cog wheels, adding rubber bands to make one wheel turn another.

Equipment: coffee-jar lids, elastic bands, corrugated paper, cotton reels, wooden board, nails, Lego Technic®, Georello®, Teko®.

Extensions: children could investigate:

- Observing if different sized cogs turn at the same rate
- Changing the direction of the cogs
- Using construction kits to make machines which use cogs.

Programmes of study: F–1a, 1b; Sc1–2b, 2c; Sc4–2a

2 Using magnets Provide the children with various objects and ask them to predict which objects will be attracted to a magnet and record their predictions before testing.

Equipment: variety of objects, pins, buttons, various magnets, paper clips, card.

Extensions: children could investigate:

- Where a magnet is strongest
- Using magnets to separate mixtures of pins and buttons, sand and iron filings
- Testing magnets through different materials such as wood, glass, plastic
- Designing and making magnetic games.

Programmes of study: F–1a, 1b, 3a; Sc1–3d, 3e, 3f; Sc3–1b

Making things fly

- Observing seeds and fruit that fall and spin
- Collecting and testing objects that can be made to fly through the air
- Playing the 'flapping fish' game
- Investigating spirals – best material, where do they spin fastest?
- Investigating ways of making a piece of paper fly
- Designing, making and testing autogyros, paper planes, parachutes, kites
- Designing and making launchers to help objects fly through the air
- Designing, making and investigating windmills. How many different ways to power, best material, size

Floating and sinking

- Early explorations, e.g. sorting collections of objects that either float or sink
- Predicting and testing objects that the children think will float, putting them into sets
- Investigating how to make milk bottle tops, Plasticine®, float by changing the shape
- Investigating how to make floaters sink and sinkers float
- Investigating 'submarines'

Balls

- Collecting a wide range of balls, discussing what balls can do
- Matching balls to purpose
- Investigating how balls bounce
- Making balls from different materials, comparing their performance to tennis balls
- Designing fair tests for measuring the bounciness of different balls
- Investigating how balls behave on different surfaces

Moving in air and in water

Balloons

- Investigating how balloons travel, comparing size and shape
- Investigating how far a balloon can stretch
- Designing and making a jet rocket using a balloon. Trying to improve its performance

Boats

- Exploring toy boats through free play in the water tray
- Designing, making and testing boats using Duplo®, Bauplay®, Plasticine®
- Investigating boat shape, sails – which travel furthest, fastest?
- Investigating boats that can carry cargo – how is the weight carried affected by the boat's shape, size?
- Designing, making and powering boats

Moving in water activities

1 Floating and sinking The children collect and test an assortment of objects which sink and float. They predict, then test to see what happens.

Equipment: sponges, toys, cork, nails, wood, paper, Plasticine®, plastic tank.

Extensions: children could investigate:

- Making floaters sink and sinkers float
- The best shape for a raft to carry cargo, e.g. broad-based, shallow sided, narrow based, tall sided.

Programmes of study: F–1a, 1c, 3a; Sc1–1c, 2a

2 Investigating boats The children use Duplo®, Bauplay®, Plasticine®, etc. to make and try out different boats.

Equipment: construction kits, plastic tank, balsa wood, wooden boat shapes, card, matchsticks, cocktail sticks, motors, elastic bands, batteries, wires, balloons.

Extensions: children could investigate:

- The best boat, exploring shape, size and shape of sail, and so on
- The amount of cargo which can be carried
- Different ways to power a boat
- Designing and making a paddleboat.

Programmes of study: F–1a, 1b; Sc1–3c; Sc3–1e; Sc4–2c

3 Making objects float and sink How can children make a straw float upright in water by adding paper clips?

Equipment: straw, paper clips, salt, plastic bottle, plastic tubing.

Extensions: children could investigate:

- Trying different solutions, e.g. adding different amounts of salt to the water and observing the effects
- Making a submarine by submerging a plastic bottle in water and forcing water out of the bottle by using plastic tubing
- Raising buried treasure using plastic tubing
- Making catamarans using squeezy bottles, adding cargo, and so on
- See also extension activities in 'Water' topic.

Programmes of study: F–1a; Sc1–1b

Moving in the air activities

1 Balls The children collect balls from around school and at home. What can balls do? (Bounce, roll, drop.) Which ball do children like best and why?

Equipment: selection of balls, metre ruler, different surfaces (e.g. carpet tiles, grass, tarmac, flags, felt, polystyrene, polythene sheets, plastic), junk material, elastic bands, cardboard tube.

Extensions: children could investigate:

- Making their own balls using fabric, elastic bands, Plasticine®, etc. They try to throw their balls or bounce them. They investigate which bounces better – a tennis ball or a home-made ball
- How high they can make balls bounce
- How many ways they can find to make a ball travel
- Sorting the balls (e.g. by size, colour, weight, bounce)
- Predicting then testing which ball will bounce best
- How many times the balls bounce before coming to a stop
- If the height at which a ball is dropped makes a difference to the number of bounces
- Devising a fair test for the best bouncing ball
- Whether the surface on which the ball is bounced makes a difference.

Programmes of study: F–1a; Sc1–1a, 1b, 2b, 2c; Sc3–1b; Sc4–2c, 2d

2 Ping pong ball launcher Design and make a ping pong ball launcher.

Equipment: junk material, elastic bands, elastic, ball, bucket.

Extensions: children could investigate:

- Fair testing
- Designing a target game using their launcher.

Programmes of study: F–1c; Sc1–1c

3 Balloons The children blow up balloons of various shapes and sizes and let them go. They then observe their flight path.

Equipment: selection of balloons, straw, card, peg, nylon fishing line, sticky tape, two posts, timer.

Extensions: children could investigate:

- How far a balloon can stretch or spring – do all balloons, whatever their shape or size, behave in the same way?
- Designing and making a rocket using a balloon. They could measure the distance travelled, and the effect of adding card/paper wings.

Programmes of study: F–1a; Sc1–2b; Sc4–2d

4 Making things fly Look at seeds and fruit that fall and spin. The children can try dropping them indoors and outdoors on a calm/windy day and note the differences.

Equipment: seeds, timer.

Programmes of study: F–1a; Sc1–2b

5 Making spirals The children make spirals from various kinds of paper and suspend them in different places. What happens? Where do they spin the fastest?

Equipment: papers of various kinds, straws, balloons, metre rule.

Extensions: children could investigate:

- Finding as many ways as possible to make a piece of paper fly – they can change the shape by folding, adding weights, and so on
- Collecting objects which can be made to fly through the air – paper, paper plates, milk-bottle tops, and so on. They could measure how far they travel
- Making a flapping fish game using tissue paper fish, and investigating how many ways they can find to make the fish move – by blowing, using straws, balloons.

Programmes of study: F–1a; Sc1–3c

6 Autogyros The children make and test autogyros of different materials (using a template so that they are all the same size). They could investigate which autogyro takes the longest time to reach the ground indoors/outdoors.

Equipment: paper of various types, card of various thicknesses, paper fasteners, paper clips, scissors.

Extensions: children could investigate.

- Slowing down an autogyro by adding paper clip weights, changing the size or material. This would be an ideal opportunity to introduce the idea of fair testing by changing only one variable at a time.

Programmes of study: F–1b; Sc1–1a, 1c

7 Making parachutes and paper planes The children could design and make paper planes and parachutes. They could then test them to see which one flies the furthest/highest/quickest, or which one stays in the air for the longest time.

Equipment: card, paper, polystyrene, polythene, weights, Plasticine®, paper clips, cotton thread, wood.

Extensions: children could investigate:

- Changing material, weight or size (remembering to change only one variable at a time)
- Designing and making gliders and boomerangs
- Designing and making launchers to help their planes/gliders travel further
- Designing and making kites.

Programmes of study: F–1a; Sc1–2c

7

KEY STAGE 2: LOWER JUNIOR TOPICS (YEARS 3 AND 4)

Clothes

Clothes for a purpose

- Make a collection of hats. Discuss the purpose they serve, who might wear them, the materials they are made of. Sort the hats in various ways. Design and make a hat to suit a given purpose
- Make an accessory to keep us safe when out at night (reflective armbands, hats)
- Clothes for special occasions, e.g. play suits
- Fastenings: speed tests, e.g. soap, candle wax
- Design an art apron that is easy to keep clean, easy to fasten and can be worn by different sized children

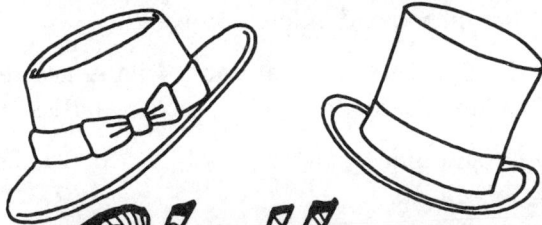

Whatever the weather

- Dressing a toy/teddy for different seasons
- Sorting clothes for keeping warm/cool, wet/dry
- Investigate which materials are best for keeping out water
- Design and make a waterproof item of clothing
- Investigate whether a glove keeps heat in – looking at different types of gloves. Compare materials made to purpose for use
- Design and make a hat that fits, is warm and dry and can be seen in the dark
- Investigate hoods
- Shoes: test for slipability. Look closely and compare different shoe soles. Make soles from different materials, e.g. sandpaper, carpet, and record the force needed to pull along a smooth surface
- Design and make an umbrella
- Which colours do we wear in hot weather? Why?

Looking after clothes

- Which soap is best for removing stains? Try hot/cold water, hot soapy/cold soapy water
- Drying times – which fabrics dry the quickest? Find the best place for drying; make a spin-drying machine
- Crease tests – which fabrics crease most/least after being scrunched and laden with weights?
- Looking at clothes labels and identifying symbols and their meanings

Fabrics

- Collecting together samples of a wide variety of fabrics, sorting according to different criteria
- Looking at weave patterns under a microscope
- Printing, dyeing fabrics
- Insulation test: investigate the best material to keep a container of water warm
- Strength tests on fibres. How much weight will a strand hold before it breaks?
- Test materials for wear and tear, e.g. rubbing with sandpaper, stone, stretching

Clothes for a purpose activities

1 For safety Ask children to design and make an accessory to keep people safe when out at night. They will need to consider how well it can be seen in the dark and how to make it waterproof.

Equipment: various materials, a home-made 'dark box' with a peephole for testing, a torch, polythene, candle.

Programmes of study: F–2a; Sc1–1c, 2b; Sc3–1a, 1e; Sc4–3a, 3c, 3d

2 Play suit Devise a test to find which fabric would be suitable for a playsuit.

Equipment: fabrics, sandpaper, pumice stone, microscope, hand lens, drawing pins, weights, strip of wood, wooden block.

Extensions: children could investigate:

● The properties of various materials, e.g. hardwearing, comfortable.

Programmes of study: Sc1–1a, 1d, 2a, 2b; Sc3–1a, 1b, 1c

Whatever the weather activities

1 The hat The children design and make a hat for themselves to keep them dry and safe at night.

Equipment: different types of fabric and materials, scissors, string, glue, paper clips, wax crayons, candle.

Extensions: children could investigate:

● Which colours can be seen at night
● Ways to make fabric waterproof.

Programmes of study: Sc1–1a, 1d, 2a, 2b; Sc3–1a, 1b, 1c

2 Hoods Investigate the effects on safety of wearing a hood.

Equipment: coats with hoods, rattle, blocks of wood, triangle.

Before this investigation, the children would need to discuss road safety (e.g. the Green Cross Code) and think about the implications of wearing a hood.

Programmes of study: F–1a; Sc1–2b, 2c; Sc3–1a

Looking after clothes activities

1 How can we remove stains?

Equipment: fabric with a stain (e.g. tomato sauce), tank/bowl, cold/warm water, cold soapy water/warm soapy water.

Extensions: children could investigate:

● How to decide which is the cleanest
● Which soap is best
● Making the test fair by thinking about the temperature, volume of water, amount of soap, washing time.

Programmes of study: F–1a; Sc1–1a, 1d, 2b, 2c; Sc3–1a

2 Crease tests Devise a test to see which fabrics crease most or least after being scrunched and laden with weights.

Equipment: natural and manufactured materials, timer, weights, junk materials.

Programmes of study: Sc1–1a, 1d, 2a, 2b; Sc3–1a, 1b, 1c

Fabrics activities

1 Examining fabrics The children make a collection of different fabrics. How are the different fabrics coloured? Use a microscope to find out more about them. In what pattern are they woven? Compare the number of threads in a square centimetre.

Equipment: various fabrics, microscope, hand lens, rule, plain paper.

Extensions: children could investigate:

● The strengths of different threads

● The effect of dyeing different materials.

Programmes of study: Sc1–1a, 1d, 2a, 2b; Sc3–1a, 1b, 1c

2 Insulation Which material will keep the water warm for the longest time?

Equipment: cans, thermometers, fabric/materials, elastic bands, warm water, measuring cylinder, funnel, timer/stop clock.

Extensions: children could investigate:

● Which is the best material to use as an insulator
● Making the test fair by using the same size of fabric sample, same volume of water, same starting temperature of water, etc.

Programmes of study: Sc1–1a, 1d, 2a, 2b; Sc3–2b

Changes

In materials

● How strong are materials? Can you make them stronger by folding, rolling, layering papers?
● What difference does treating materials make, e.g. by wetting or oiling?
● Explore what happens when kitchen substances are mixed together. How does cooking affect this?
● How do building materials change over time, e.g. weathering, cement setting?
● Separating mixtures

In ourselves

● Collect a data bank of individual measurements, e.g. height, reach, stride. What links and patterns can you find? Are they the same for younger children?
● How do we change after exercise, e.g. colour, breathing rate, pulse rate?

In directions

● How can you bring about changes in the direction of a moving object, e.g. using cogs, gears and pulleys?
● Can you change spinning directions, e.g. autogyros, motors, propellors?
● How does changing shape affect direction, e.g. flaps on gliders?
● Mapping pathways, routes for children, toy cars, computer controlled vehicles
● Look at the changes that occur in the directions that magnets, pairs of magnets can point to – using compasses
● Using mirrors

In the environment

● Adopt a patch (a tree, a piece of school ground, part of a park). Observe and record its changes over the school year.
● How do living things change in relation to changes in the environment, e.g. birds migrating, camouflage, leaves changing colour?
● How does the growth of plants change under different conditions?
● Observe and record the changes in the life cycles of small creatures, e.g. stick insects
● What effects do changes in the speed and direction of wind have?
● Investigate changes in day length over the year

Changes in materials activities

1 Strengthening paper How can you make paper stronger?

Equipment: newspaper, sticky tape, glue, weights, cane/broom handle.

Extensions: children could investigate:

- Building structures and testing their strength
- Looking at a variety of papers and materials and considering ways of making them stronger, for example by treating their surfaces.

Programmes of study: F–1a; Sc1–1a, 1b, 2a, 2b; Sc4–2f

2 Bubble mixtures Find mixtures which make bubbles.

Equipment: vinegar, egg shells, sugar, bicarbonate of soda, Andrews® Liver Salts, water, yeast, plastic containers, milk bottles, balloons, wine bubblers.

Extensions: children could investigate:

- The combinations which make bubbles
- Varying the quantities of different substances and finding ways to measure how quickly the bubbles are produced (e.g. using balloons stretched over the necks of bottles, using wine bubblers)
- Which mixture is best for blowing bubbles
- What difference the shape of a bubble-blower makes.

Programmes of study: F–1a; Sc1–1a, 1d, 2a, 2b; Sc4–2f

3 Separating things How can you separate mixtures?

Equipment: dry sand, sugar, magnets, rice, iron filings, salt, lentils, water, funnels, sieves, filter papers, foil, containers, timer.

Extensions: children could investigate:

- Ways to separate combinations of materials
- Making problem mixtures for others to separate
- Setting a time limit or, alternatively, making the investigation into a race – who can sort out the materials in the fastest time?

Programmes of study: F–1a; Sc1–1a, 1d, 2a, 2b

Changes in ourselves activities

1 Looking for links and patterns Collect a data bank of information about the children on height, weight, shoe size. Ask them to look for links and patterns.

Equipment: bathroom scales, metre rules, access to a computer database programme, timer.

Extensions: children could investigate:

- Collecting information from older and younger children
- Devising fitness tests for themselves
- Drawing up time-lines of their day. When are they most active? How much time do they spend on various activities?

Programmes of study: F–1a; Sc1–1a, 1d, 2a, 2b; Sc2–4a

Changes in direction

1 Autogyros Investigate which material makes the best spinner. Can children make one which spins more quickly or spins in the opposite direction?

Equipment: a variety of papers, scissors, paper clips, rulers, pencils.

Extensions: children could investigate:

- Adding paper clips
- Damaging one flap.

Programmes of study: F–1a; Sc1–1a, 1d, 2a, 2b; Sc4–2h

2 Magnets and compasses Where does a compass point? How can children make it change direction?

Equipment: compasses, magnets, a variety of metallic/non-metallic materials, string, paper.

Extensions: children could investigate:

- Suspending a magnet or using a compass to find out what happens when a variety of materials, including other magnets, are brought near to it
- How far away one magnet can influence another
- The strengths of different magnets.

Programmes of study: F–1a; Sc1–1a, 1d, 2a, 2b; Sc4–2a

3 Using mirrors to make a kaleidoscope Using plastic mirrors hinged together with sticky tape, and a selection of small objects such as Unifix or coloured beads, investigate the relationship between the number of images and the size of the angle between the mirrors.

Using this information, can children think how kaleidoscopes work?

They then design and make a kaleidoscope and make observational drawings of what happens to the pattern when they change the angle of the mirrors. How many patterns can they make?

Equipment: hinged plastic mirrors, coloured beads, Lego®, Unifix, paper, protractor.

Extensions: children could investigate:

- The path the light takes in a kaleidoscope
- Making periscopes.

Programmes of study: F–1a; Sc1–1a, 1d, 2a, 2b; Sc4–3b, 3c

Changes in the environment activities

1 Around school Adopt a patch of land near school. Observe and record its changes over a school year.

Equipment: camera, paper, seeds, different growing mediums, minibeasts, binocular microscope, hand lens.

Extensions: children could investigate:

- Growing plants under different conditions
- Observing and recording the changes in the life cycles of small creatures.

Programmes of study: F–1a; Sc1–1a, 1d, 2a, 2b; Sc2–1a, 1b

2 Day and night Observe with care the position of the sun at intervals during the day. Compare shadow size and shape.

Equipment: simple measuring instruments, star charts, relevant books.

Extensions: children could investigate:

- Records of day length, using newspaper information
- Records of the shape and size of the moon, the positions of the stars, and so on
- Detailed weather records.

Programmes of study: F–1a; Sc1–1a, 1d, 2a, 2b; Sc4–4d

Communications

Ourselves

- By touching: using feely bags, touch test
- Hot and cold: body temperatures
- Using sight: eye tests for colour and distance
- Listening: pinpointing sounds, listening test, muffling sounds
- Using smell: strong, lingering smells, how far do smells spread?
- Memory games: Kim's game, sorting by colour, shape, speed reaction tests
- Our bodies: pulse rates, breathing rates

Signs and symbols

- Identifying colours which stand out at a distance or in the dark
- Colour as a warning, going on a colour walk to identify road signs
- Food colouring: how important is it?
- Testing colours, dyeing, finding colour, separating colour
- Colour in nature – as a warning signal to others, for camouflage

Devices which help

- Telephones: making, testing and improving string telephones
- Electricity: morse code, alarm systems, traffic lights
- Using mirrors: investigating security mirrors in shops, making periscopes
- Collecting and sorting devices which help us to see and hear better: ear trumpets, megaphones, stethoscopes, binoculars, spectacles, telescopes

Using systems

- Sending messages – investigating papers, pens, making and using invisible ink
- Sending coded messages using a torch, a buzzer, devising codes
- Investigating the postal system, addresses, post codes, street plans
- Newspapers: how they are made, comparing contents, reports of different events, making a class newspaper

Ourselves activities

1 What can we find out by touch Use feely bags to test how difficult it is to identify objects by touch alone. Start by including a wide variety of odds and ends (cold spaghetti to which a drop of oil has been added always provokes a reaction!) and graduate to collections of similar objects, e.g. coins, fabrics, papers. Objects can be described without naming them – children can make drawings of the objects and then compare their drawing with the contents.

Equipment: feely bags, a variety of objects, paper, thermometer, water at different temperatures.

Extensions: children could investigate:

- Sorting more specific items, e.g. grading sand paper, identifying papers and cards by thickness or texture

- How wearing different gloves affects their sense of touch
- Using other parts of their bodies, for example their toes
- Where their skin is most sensitive to touch
- How good they are at identifying hot and cold water
- Measuring the temperature in different parts of the classroom
- Measuring body temperature
- Designing and making a poster which could be identified by a blind or by a partially-sighted person.

Programmes of study: F–1a; Sc1–1a, 1d; Sc2–4a; Sc3–1a, 1b

2 Using memory The children play Kim's game by putting a number of objects on a tray, giving other children a certain amount of time to look at the tray and then seeing how many objects they can remember. They could design and test 'Spot the difference' pictures (e.g. photographs of the school ground taken at different times of the year).

Equipment: tray, objects, card, timer.

Extension: children could investigate:

- Devising word and number tests
- Trying the tests with added distractions – e.g. in the playground, while playing music – to see if this affects the ability to remember
- Reaction times using rulers, testing foot reaction time, etc.

Programmes of study: F–1a; Sc1–1a, 1b, 1c, 1d; Sc2–4a

Signs and symbols activities

1 Colour around us Ask the children to list all the ways in which colour helps at home, in school, in the street, in shops, on the roads. They could then classify using criteria such as helpful, warnings, for effect.

They could then devise sorting tests to see if people are quicker when they sort by colour or by shape.

Equipment: coloured objects, dark box, timer, coloured papers, metre rule.

Extensions: children could investigate:

- Making their tests fair
- Which colours stand out at a distance, in the dark, and use this knowledge to make posters, design road traffic warning or information signs.

Programmes of study: F–1b; Sc1–1a, 1d; Sc4–3b, 3c

2 Testing colours Ask the children to add food colouring or paint to water, drop by drop, and to observe what happens. They should try with different colours, adding the colour from different heights and then noting the effect. They could then try diluting colour by adding a specific number of drops of colour to water and systematically diluting the colour by pouring half of the solution into another container which they top up with water.

How many times can they dilute and still see the colour?

Equipment: food colouring, water, plastic containers, Smarties®, coloured felt tip pens, a variety of black felt tip pens, filter papers, droppers.

Extensions: children could investigate:

- Making shade cards using paints
- Separating colour using Smarties®, felt tip pens (make sure that the pens are water soluble)
- Different black felt pens
- Dyeing fabric using natural material and commercial dyes, using mordants.

Programmes of study: F–1a; Sc1–2b

71

Devices that help activities

1 String telephones The children work with partners to make the best telephone.

Equipment: a selection of pairs of containers with holes in the end (e.g. boxes, yoghurt pots, cans, plastic or polystyrene cups), strings, twine (of various lengths and thicknesses), weights, wax, nails, water.

Extensions: children could investigate:

- The effect on the sound transmission of the length or thickness of the string
- The effect of hanging weights on or treating the string
- The need for a standard 'sound' to test their various models.

Programmes of study: F–1a; Sc1–1a; Sc4–3e, 3f

2 Transmitting messages Using simple and more complex electric circuits, the children design and make devices to transmit messages. These could include a morse code transmitter, an alarm system using a bulb or buzzer, a simple torch and a set of traffic lights.

Equipment: bulbs, batteries, wire, buzzer, tools, junk material, switches.

Programmes of study: F–1a; Sc1–1a

Using systems activities

1 Sending letters The children test different kinds of paper in various ways, matching type to purpose.

They investigate the best way to send a secret message using lemon juice and other substances.

Equipment: a variety of different papers, paints, crayons, pencils, lemon juice, orange juice, milk, bicarbonate of soda.

Extension: children could investigate:

- Strength tests on paper
- The postal system and how post codes are used
- Packaging for different purposes.

Programmes of study: F–1a; Sc1–1d

2 Coded messages The children list different ways of sending a coded message. They should consider the recipient, the distance to be covered and how to keep it secret. They then use the materials provided to send messages.

Equipment: torches, mirrors, electricity equipment, paper (for inventing codes), card (for megaphones, and so on).

Programmes of study: F–1a; Sc1–1a; Sc4–1a, 1b, 3c

Making a vehicle

- Using a variety of techniques: Jinks®, Lego Technic®, Capsela®, Teko®, junk
- To move on land, in or on water, in air, through space
- Make a boat, rocket, car, glider, sledge, land yacht

How can you make it move?

- Can you use your own 'body power' to make a vehicle move, e.g. blowing, pushing, pulling, throwing?
- Which method is most appropriate? Which is easiest?
- How else could you power a vehicle, e.g. using elastic bands, balloons, fans, bellows, propellers, gravity, paddles, motors?

Making it move better

- How does the material that a vehicle is made from affect how it moves?
- What difference does weight make?
- How does friction help or hinder?
- What difference does changing friction make, e.g. lubricating parts, changing surfaces of tyres, roads
- How does shape affect movement (streamlining, aerofoils, drag)?
- How can cogs, gears, pulleys, ramps, be used to make movement easier, or alter the movement?
- Investigate hydraulics

Transport and safety

- Investigate safety measures, e.g. seat belts, helmets, protective clothing, and compare for strength in relation to impact, sliding
- Explore stopping devices – crash barriers, sand, foam, parachutes, braking mechanisms
- How is stability affected by shape, load, position of load?
- Devise quick exits from a train, boat, plane: chutes, ejection seats

Making a vehicle activities

1 Moving on land Ask the children to design and make a land vehicle which can carry a load and with which they can investigate the size of load which it can carry. Does the surface on which it is travelling affect the speed, the load or the stopping distance?

Equipment: Lego Technic®, Teko®, Capsela®, dowel, cardboard triangles, scissors, glue, rulers, wood, saws, card, various wheels.

Extensions: children could investigate:

- The stability/load relationship
- What their vehicle could be used for
- If wheel size makes a difference to the vehicle's speed or the way it moves.

Programmes of study: Sc1–1a, 1d, 2a, 2b; Sc4–2h

2 Moving through the air Design and make a parachute which will fall as slowly as possible to the ground.

Equipment: variety of fabric/paper, weights, Plasticine®, string, scissors.

Extensions: children could investigate:

- Using different sizes of canopy
- Using different lengths of string
- Using different weights
- How slowly their parachute falls to the ground
- If the test is fair.

Programmes of study: Sc1–1a, 1d, 2a, 2b, 2c; Sc4–2b

3 Moving in water How many different ways can children find to make a 50 g ball of Plasticine® float?

Equipment: Plasticine®, water, junk, weighing machine, tank, salt.

Extensions: children could investigate:

- Changing the shape of the Plasticine®
- Adding different amounts of salt to the water.

Programmes of study: F–1a; Sc1–1a; Sc4–2g

Making it move activities

1 Powering a vehicle Ask children to find as many different ways as they can to make a vehicle move without touching it.

Equipment: a simple basic chassis (made using Teko® or Jinks' technique), a motor, a ramp, string, gears, batteries, bellows, elastic, pulleys, cardboard, wires, rubber bands, weights, balloons, junk.

Extensions: children could investigate:

- How far will their vehicle can go
- Making it go in the opposite direction
- Making it go over different surfaces, up a slope, at a different speed.

Programmes of study: Sc1–1a, 1d, 2a, 2b; Sc4–2f, 2g, 2h

Making it move better activities

1 Cogs, gears and drive-belts How many ways can children find for making one wheel turn another?

Equipment: cotton reels, different sized cog wheels, elastic bands (a variety of sizes and thicknesses), nails, wooden base or shoe box lid, motors, batteries, wires.

Extensions: children could investigate:

- Making more than one cog turn
- If a larger wheel joined to a smaller one goes round faster or slower
- Whether the thickness of the elastic band affects performance
- Designing and making their own cog wheels.

Programmes of study: Sc1–1a; Sc4–2h

2 Hydraulics Children use a balloon as a hydraulic lift to empty a truck.

Equipment: plastic tipper truck, balloon and balloon pump, Plasticine®, dried peas, syringe/plastic tubing.

Extensions: children could investigate:

- Different ways of using a balloon
- If changing the load makes a difference
- The relationship between the amount of air in the balloon and the height of the 'tipper' on the truck.

Programmes of study: Sc1–1a, 1d; Sc4–2h

Transport and safety activities

1 Seat belts Children design and make a seat belt which will keep Snoopy safely on the sledge when he comes to the bottom of the ramp.

Equipment: strapping, Snoopy or other soft toy, toy sledge, slope, elastic.

Extensions: children could investigate:

- Finding the best material for a seat belt
- Wearability tests on different materials
- Changing the angle of the slope
- Testing a wheeled vehicle, altering the surface.

Programmes of study: Sc1–1d; Sc4–2h

2 Safety chutes Design and make a safety chute for a plane. Children investigate the angle of the slope, the surface on landing.

Equipment: junk material, card, different surfaces.

Programmes of study: F–1a; Sc1–1a

Supermarket

Containers

- Sorting packages to specific criteria/own criteria
- Find the best way to pack a box using different shaped items
- Which foods are packed in tins?
- Collecting labels together: looking at weight, ingredients, place or origin, direction
- Size and shape of containers: use one piece of card, find which shape holds the most
- Collect together carrier bags – test to find out which carries the most weight
- Design and make your own carrier bag that is strong, attractive and comfortable to hold
- Design and make a model shopping trolley using Lego® and junk: can it be adapted for a wheelchair?

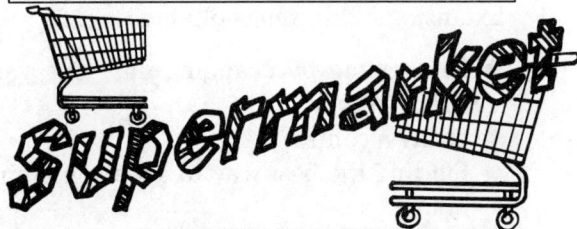

Food

- Collect together a variety of cereals. Carry out taste tests, surveys, mixtures of cereals to find ways of separating them
- Fruit: cutting, drawing, printing, weighing, rolling
- Insulation test: find the best material to keep the ice cube the longest
- How long do frozen things stay frozen?
- Ice cube in different places: observe and record
- Note symbols on frozen packages and freezers: what do they mean?
- Sorting foods into solids and liquids: if you can pour it is it necessarily a liquid, e.g. sugar?
- What happens when chocolate, butter, jelly, are placed in foil boats on warm water, then in cold water? What changes occur?
- Investigate everyday ingredients, e.g. coffee, salt. What happpens when poured into hot water/cold water, stirred/unstirred?
- Cooking – following a recipe, noting changes to ingredients before/after cooking

The environment

- Make a burglar alarm to protect the supermarket premises
- Design and make a model for a collecting site for recycling materials
- What products can be recycled?
- Carry out a survey to find out which products use recycled materials
- Sort a variety of materials (e.g. polythenes, papers, plastic card) into biodegradable/non-biodegradables
- Design and make a magnetic tin sorter
- Make your own paper. Try a newspaper, non-laminated card, paper bags, paper towels

Displays

- Flashing sign advertising the supermarket
- Visit to supermarket and find out reasons for the layout
- Colour: which are the colours we notice first? Design a colour test
- Design and make a sign to show where a specific food is, for a blind person, someone who does not speak English
- Design and make a coin testing machine
- Find out the most stable pattern for stacking tins, packets or biscuits, bars of Toblerone®
- Design and make an attractive package for your own product

Containers activities

1 Different shaped containers Using one piece of A4 card each time, children make as many different shaped containers as they can. Which shape has the largest volume?

Equipment: sheets of A4 card, sticky tape, scissors, marbles.

Extensions: children could investigate:

- Shapes which are best for stacking
- Which goods are packed in tins, and suggest why
- The labels of tins and packets and what they tell us.

Programmes of study: F–1c; Sc1–1b

2 Packaging Sort the containers into sets of different materials. How many different ways can children sort them? Work with a partner: one sorts and then the partner guesses the criteria used.

Equipment: empty food packages, plastic bottles, foil containers, frozen food bags.

Extensions: children could investigate:

- Which packages are suitable for liquids and why
- Making containers which are suitable for liquids
- Designing an eye-catching package for a product.

Programmes of study: F–2a; Sc1–1a

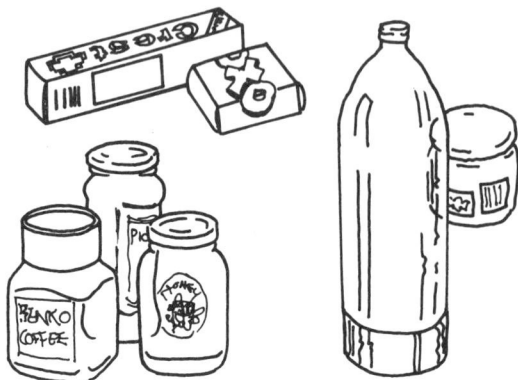

Food activities

1 Cereals What is the favourite cereal in the class? Conduct a survey to find out.

Equipment: variety pack of cereals, empty cereal packets, measuring cylinders, sieves, colanders, plastic nets (the kind that fruit is packed in), card, scissors, squared paper.

Extensions: children could investigate:

- Ways of separating mixtures of cereals
- Comparing the volumes of the different cereals
- The amount of information on each packet
- Which cereal is the best value for money (comparing weight and price).

Programmes of study: F–1c; Sc3–3a

2 How long do frozen things stay frozen? Children devise ways of keeping an ice cube. Try in different places, using different materials. Try to keep the test fair.

Equipment: ice cubes of a uniform size, Petri dishes, straw, newspaper, polythene, fabric, other materials suggested by the children, timer.

Extensions: children could investigate:

- Fair testing by changing only one variable at a time
- Using a control
- Finding the best way to keep something hot.

Programmes of study: Sc1–1d; Sc3–2b, 2d

The environment activities

1 A burglar alarm Design and make a burglar alarm to protect supermarket premises.

Equipment: batteries, bulbs, wires, buzzer, switch, drawing pins.

Programmes of study: F–1a; Sc1–1a; Sc4–1a, 1b, 1c, 1d

2 Recycling Design and make a model of an area for the collection of materials that can be recycled.

Equipment: junk, magnets, scissors, card, glue, pens, string.

Extensions: children could investigate:

- Which materials can be recycled
- Which type of containers would keep the area safe, clean and tidy.

Programmes of study: F–1a; Sc3–2c

Displays activities

1 Signs Design and make an illuminated sign for a supermarket. Can children make it flash?

Equipment: batteries, wires, bulbs/bulb holders, switches, buzzer, screwdrivers, drawing pins, paper clips, wooden block, card, cellophane, glue, scissors, felt tips.

Extensions: children could investigate:

- Constructing a simple circuit to light a bulb
- Making a switch to add to the circuit
- Replacing the bulb with a buzzer.

Programmes of study: F–1a; Sc1–1a; Sc4–1a, 1b, 1c

2 Colour Colour is important in packaging. Ask children to devise a test to determine whether one colour is easier to remember than another.

Equipment: squared paper, timer, shapes in different colours, different sizes.

Extensions: children could investigate:

- The use of different colours for posters/ packaging
- Devising eye tests to see if some people can see some colours better than others
- Colour-blindness tests
- If sorting by colour is easier than sorting by shape.

Programmes of study: Sc1–1d

On a desert island

Communication

- Design and make a watch tower using a limited amount of straws and sticky tape
- Devise an alarm to warn off intruders, using a pressure pad
- Keep a diary of events, observations, weather changes and discoveries on your desert island
- Design and make a sundial
- How many ways can you find to send a message, e.g. message in a bottle, flashing lights using electricity, mirror reflecting light, using flags, Morse code using a buzzer
- Design and make a kite

Leisure

- Draw a map of your island, e.g. vegetable plot. Using coordinates, plan the areas. Devise a route for other children to follow to find their way round the island
- Buried treasure – use coordinates to find it, or use a magnet
- Devise a key showing all the characteristics of the flora and fauna inhabiting the island
- Sorting shells, coral, pebbles, feathers and other natural materials. Think of different ways of recording
- Design and make some games using junk materials and devise rules to play

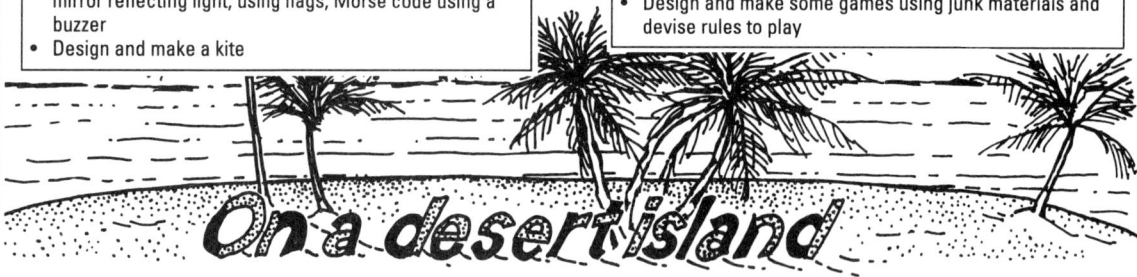

On a desert island

Food, clothes, shelter

- Fruits and vegetables: sorting and classifying, painting and dyeing
- Growing food/seeds in different conditions and record differences in growth
- Design and make a utensil to collect water. How can you purify the water?
- Devise a fair test to find the best way to keep the water cool/warm
- Design and make a shelter. Think about the best siting; the areas you will need within the shelter; the best materials to make the shelter
- Can you design and make a hammock to hold a certain weight?
- Waterproofing, sun proofing materials to make them more durable

Transport

- How many ways can you find to move your raft, e.g. elastic band power, sails, electric motors?
- Moving objects around the island. How many ways can you find to move a load, e.g. cotton reels, slope, pulleys?
- Using Lego Technic®, design and make a buggy to transport you around the island
- Explore the different ways you can find to lift the treasure chest up to your tree shelter, e.g. use a Lego Technic® crane, pulleys and levers
- Design and make a bridge to span a certain distance and carry a weight

Communication activities

1 Sending messages Send a message to another group of people on the island.

Equipment: bulbs, bulb-holders, batteries, wires, paper clips, coloured cellophane, card, silver foil, mirrors, buzzers.

Extensions: children could investigate:

- Ways of using a simple circuit to communicate with a buzzer or a light
- Reflecting the light using a mirror
- Devising codes.

Programmes of study: Sc1–1d; Sc4–1a, 1b

2 More messages... Children explore the number of ways they can find to send a message. Can they identify an effective way of attracting attention from another island, or from the mainland, or from a passing ship or aeroplane?

Equipment: bulbs, batteries, cellophane, card, wood, polythene, tins, elastic bands, mirrors, string, tissue, cotton, pens, plastic bottles, greaseproof paper, balloons.

Extensions: children could investigate:

- Making a megaphone using card and junk materials
- Designing and using flags
- Which is the best pen for sending a message in a bottle
- Using simple circuits and switches
- Making kites
- Making patterns using sand and shells
- Using a mirror/series of mirrors to reflect light.

Programmes of study: Sc1–1a; Sc4–1a

Leisure activities

1 Exploring the island Sort and classify the natural materials. The children should find different ways to record their results.

Equipment: shells, pebbles, coral, feathers, fir cones, bark, leaves, flowers, seeds.

Extensions: children could investigate:

- Different ways of sorting and asking other children to guess the criteria.

Programmes of study: F–1a; Sc1–1b, 1d; Sc3–1d

2 Helping others to explore Sort and classify the plants and animals living on the island. Can children give each plant/animal a name and devise a key?

Equipment: pictures/drawings of four 'invented' plants (all slightly different, e.g. spiky leaves, no flowers, different shaped leaves), pictures/drawings of four 'invented' animals, clue books, squared paper. Science fiction books and comics are good sources for weird plants and creatures!

Extensions: children could investigate:

- Classifying plants and animals to a range of criteria
- Making a key for identification
- Naming the species
- Finding ways to transfer the information on to a computer.

Programmes of study: Sc2–4a

Food, clothes and shelter activities

1 Using the materials available Ask the children to think and plan what they would need to live on a desert island. How could they use these materials to make the things they need?

Equipment: variety of junk materials, plastic, cellophane, elastic bands, string, wood, fabrics.

Extensions: children could investigate:

- Making items from materials provided to use as clothing, shelter, for food collection, preparation.

Programmes of study: Sc1–1a; Sc3–1b

2 Improving materials Choose a material. How would children make it waterproof or stronger and adapt it for various uses?

Equipment: newspaper, variety of papers, scraps of fabric, sticky tape, string, polythene, candles.

Extensions: children could investigate:

- Using materials to make a waterproof or sunproof hat
- Making suitable footwear
- Adapting materials to make a shelter
- Using materials to span a gap.

Programmes of study: Sc1–1a; Sc3–1a

Transport activities

1 Moving things Children investigate the number of ways they can find to move a treasure chest along the beach to the bottom of the cliff.

Equipment: 'treasure' in a box, pencils or dowel, string, marbles, tin lids, construction sets with wheels (e.g. Teko®).

Extensions: children could investigate:

- Using string to pull the box
- Making carts from the construction toys
- Using marbles and tin lids to move the box
- Using pencils or dowel as rollers.

Programmes of study: F–1b; Sc1–1a

2 Moving things Children explore the number of ways they can lift the treasure chest up from the beach to the cliff top.

Equipment: Lego Technic®, Teko®, pulleys, string.

Extensions: children could investigate:

- Making a crane using the construction kits
- Using pulleys and levers to raise the chest.

Programmes of study: Sc1–1a, 1d; Sc4–2f, 2h

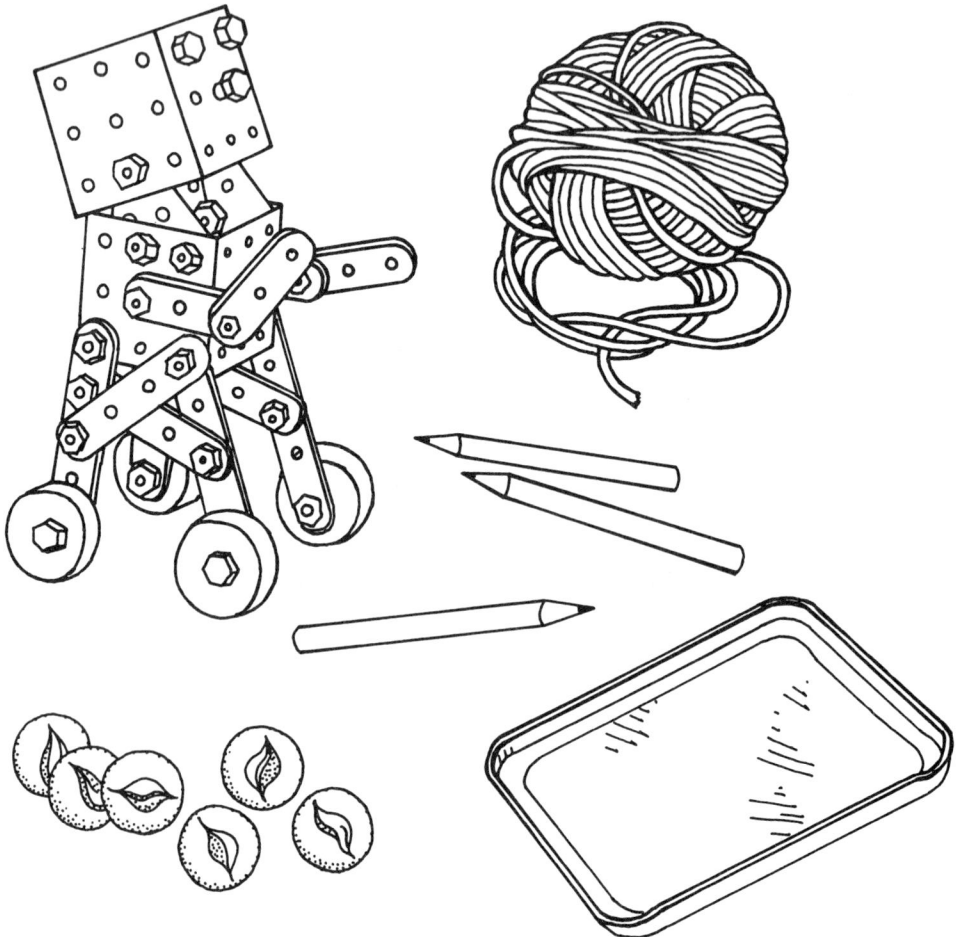

8

KEY STAGE 2: UPPER JUNIOR TOPICS (YEARS 5 AND 6)

All around our school

Playing fields/ garden

- Sample and compare soil from different locations, e.g. by wall, under trees
- Find ways of separating soil. Use a soil testing kit to identify type of soil.
- Set up a weather station to measure rainfall, wind power and temperature
- Observe body changes after exercise, e.g. pulse rate, body temperature
- Measure length/width of playing fields
- Make a survey of a marked area of the field. Record findings
- Write clues and devise a route for a treasure hunt

The playground

- Compare and test different playing surfaces for safety
- Look at structures in the playground, e.g. climbing frames
- Make stable structures
- Design an ideal playground and devise rules
- Look at suitability of footwear for playing in
- Do a litter survey. Sort and classify. Collect samples safely
- Devise games for the playground. Consider rules and safety. Make a book about children's games past and present

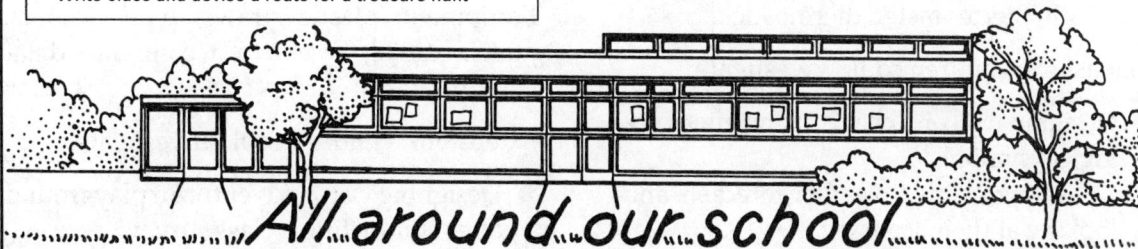

All around our school

Animals and plants

- Map out a nature trail for another class to follow
- Go on a minibeast hunt. Observe minibeasts in own habitat
- Find ways of sorting and classifying them and devise own key
- Make your own wormery. Observe and interpret findings
- Set up a bird table and observation station
- Do a survey of a variety of plants/trees and their living conditions
- Set up a 'wild' area. Create habitats. Devise your own key using stones and old planks. Make signs and labels to mark this area

Buildings

- Design and make security devices or signs for around school, e.g. burglar alarm
- Take rubbings of walls, bark
- Draw a plan of the school. Devise a plan for a visitor to show them around the school
- Devise ways of improving environment, e.g. collection of litter, quiet areas
- Do a traffic survey. Look at road markings, signs and symbols
- Find out about the history of the school
- Collect photographs and old log books
- Bury a box containing items which represent school life. How can you protect its contents?

Playing fields/school garden activities

1 Soil testing Ask children to collect soil samples from different locations, such as garden and school field, and then sort and classify according to contents of soil.

Equipment: soil samples, sieves, containers, hand lenses, microscope, disposable gloves, water, filter, paper.

Extensions: children could investigate:

- How the soil is made us (e.g. bark, stones, clay)
- The type of soil
- Separating the soil by filtering, sieving.

Programmes of study: Sc1–2a; Sc3–1d

2 Weather station Set up a weather station in the school grounds. Children devise instruments to measure rainfall, temperature, wind direction and speed.

Equipment: measuring cylinder, funnel, old tights or lightweight material, wire, canes, max/min thermometer, thermostick.

Extensions: children could investigate:

- Designing record charts using meteorological symbols
- Collecting printed weather forecasts and looking at their accuracy
- Making a quiz board of symbols used in television forecasts
- Designing weather vanes.

Programmes of study: F–2a, 2b; Sc1–1c, 3b

Playground activities

1 Playing safely Investigate which surface is best for a playground. Children devise tests for safety, bounciness and slipperiness.

Equipment: variety of surfaces (e.g. gravel, peat, tiles), variety of balls, Plasticine®, rulers.

Extensions: children could investigate:

- Bouncing different sized and shaped balls and measuring the height of the bounce
- Pulling shoes along a board covered with different surfaces and measuring the force (using a Newton meter)
- the effects on Plasticine® of dropping it on to different surfaces
- the design of an ideal playground (with apparatus, special areas) and rules for safe play.

Programmes of study: F–4b; Sc1–1a; Sc4–2c

2 Playground structures The children design and make models of structures which could be used in the playground. They then test them for stability and consider the materials they would use if the structures were outside.

Equipment: plastic straws, pipe cleaners, junk material, access to a computer database program.

Extensions: children could investigate:

- Designing an adventure playground after conducting a class survey
- Devising playground games.

Programmes of study: F–2b, 2c; Sc1–1c

Activities with animals and plants

1 Nature trails and maps Ask children to make a map of the school ground which identifies the main buildings, grassy areas, trees and flowers and tarmac areas. They should use measurements to make the map accurate.

Equipment: squared paper, appropriate measuring instruments.

Extensions: children could investigate:

- The similarities and differences between their plan and others'
- Designing a nature trail of the school grounds for younger children, using symbols
- Producing a nature trail of the area around school, considering what they will ask people to notice, how they will make sure that people understand their instructions.

Programmes of study: Sc1–1a

2 Plant and tree survey Make an autumn collection of leaves. The children can classify the leaves using a key. They should then decide how to make a display of their leaves.

Equipment: collecting bag, hand lens, relevant books, water.

Extensions: children could investigate:

- A way of measuring the dirt found on the leaves
- Damaged leaves and try to identify the cause
- The life cycle of one tree.

Programmes of study: F–1a, 1b; Sc1–2b, 2c; Sc2–2a, 2b

Buildings activities

1 What do we throw away? Investigate the contents of a class rubbish bin. (Rubbish should be collected and sorted beforehand by you, paying particular attention to the health and safety aspect. When children are sorting rubbish they must wear disposable gloves.)

Equipment: bin with a variety of contents (e.g. peel, pencil shavings, paper, crisp bags, pens, straws, old books), disposable gloves.

Extensions: children could investigate:

- Sorting and classifying contents
- Biodegradable and non-biodegradable materials
- Possibilities for recycling
- Systems for rubbish collection, then designing and making a rubbish bin for the classroom or playground.

Programmes of study: Sc1–1a; Sc2–2e; Sc3–1a

2 Warning signals Use simple circuits to make signs and warnings around the school.

Equipment: wires and crocodile clips, buzzers, bulbs and bulb-holders, batteries, junk materials.

Extensions: children could investigate:

- Making signs for around school (e.g. an illuminated 'Walk' sign for the corridor)
- Designing and making a sign or buzzer to give permission to enter a room
- Burglar alarms for school, to protect the computer or video, using pressure pads.

Programmes of study: Sc4–1a, 1b, 1c

Lights, sound, action!

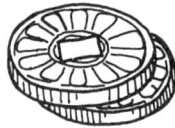

Lights

- Collection of objects/torches/mirrors. Play, sort and classify
- Colour filters: how do colours change?
- How can you change the size/shape of shadows?
- Make footlights for show
- Switch on a light with a pressure pad
- Use coloured light signals to signal the start of a play

Action

- Can you make illusions with mirrors?
- Collection of home made, manufactured puppets
- Make a shadow puppet or a puppet with moving parts
- Use pulleys, levers or gears to move scenery
- Use magnets to make parts or scenery move
- Design and make costumes or masks
- Write an action plan for a play. Consider the timing and sequencing of events

Sound

- Listening to, making, identifying sounds
- Making sound effects/musical background using *Compose*, home made or manufactured instruments
- Taping sounds, e.g. footsteps, doors opening – using them to accompany a drama
- How can you make sounds louder, quieter?
- Make a buzzer to sound an alarm
- Investigating sound vibrations

Activities with theatre lights

1 Make a set of lights Equipment: bulbs, batteries, bulb-holders, wires of varying lengths, switches, acetate sheets.

Extensions: children could:

- Find a way to switch bulbs on independently and in sequence
- Make a set of lights to signal the start of a performance.

Programmes of study: Sc4–1a, 1b, 1c, 1d

2 Flashing lights Using electricity, design and make a theatre sign that flashes on and off.

Equipment: batteries, bulbs, wires, screwdrivers, coloured acetate/cellophane, card, felt tips, scissors, paper clips, wood, drawing pins.

84

Extensions. children could investigate:

- How to light up their model theatre (e.g. spotlights, footlights)
- How to make switches.

Programmes of study: Sc4–1b, 1c

3 Coloured filters Look at a variety of colours through the filters. How do the colours change?

Equipment: bulbs, torches, coloured acetate/cellophane, different coloured papers, card, felt tips, scissors.

Extensions: children could investigate:

- How coloured filters over the lights affect the mood of the scenery
- Mixing different coloured filters and then designing a poster where different features appear or disappear with different filters.

Programmes of study: F–1a; Sc4–3a, 3c

Activities with sound

1 Composing music Use the computer program *Compose* (for the BBC B) to create pieces of music for different times throughout a performance, e.g. the overture, the intermission, the finale.

Equipment: computer, *Compose* program, tape recorder, paper, percussion instruments.

Extensions: children could investigate:

- Ways of making different sound effects to create mood changes
- Writing their own tunes and playing them on instruments which they have made
- Consider the suitability of their music for an overture, accompaniment or finale

- Use junk materials to create sound effects.

Programmes of study: Sc4–3e, 3f

2 Muffling sounds Ask the children to design and make a pair of ear muffs that will protect their ears from loud noises.

Equipment: variety of materials, e.g. fabrics, padding, yoghurt pots, string, tape measure.

Extensions: children could investigate:

- The sound-muffling properties of various materials
- How sound travels through different materials
- Devising a fair test to find the most efficient ear muffs, applying their knowledge of materials.

Programmes of study: Sc4–3g

3 Sound vibrations Investigate how volume affects vibrations.

Equipment: battery operated tape recorder (not mains) and music. It is useful to have music with a strong bass line, or a tape recorder with a bass control. Different containers to cover speaker (e.g. margarine tub, tin, wooden boxes), variety of covering materials (e.g. clingfilm, tights, balloon), variety of 'bouncers' (e.g. rice, lentils, pasta, dried peas). It would probably be advisable for you to set up a tape recorder with containers placed on top of the speaker and limit the 'bouncers' to prevent accidents!

Extensions: children could investigate:

- Altering the volume level and noting the effect
- Trying different sound effects.

Programmes of study: Sc4–3e, 3f

4 Tuning forks Using a tuning fork, children investigate the materials provided to see how sound is made by vibrations.

Equipment: bowl of water, empty bowl/jars/containers, materials to stretch over top (e.g. tights, greaseproof paper, stretched balloons, clingfilm), rice, lentils, sand, pasta, dried peas, tuning forks, elastic bands.

Extensions: children could investigate:

- Classifying the sounds by pitch
- Noting the rate of vibrations and relate this to pitch and amplitude of sound
- Making their own drums, trying out a variety of containers (deep/shallow, wide/thin) and a variety of surfaces covering the drum (stretched/loose, thick/thin) and observing the vibrations made using rice, peas, and so on
- The ripples made by the tuning fork in water. They can be encouraged to explain to others how sounds are made.

Programmes of study: Sc4–3f, 3g

Action activities

1 Mirrors Investigate ways of making figures and objects disappear using hinged mirrors. What difference does changing the angle of the mirrors make to the number of images seen?

Equipment: pairs of hinged mirrors, objects, figures, rulers, protractors, sheets from *Science for Children with Learning Difficulties* (Macdonald Education).

Extensions: children could investigate:

- The number of images seen when the mirrors are placed at different angles relative to each other
- The relationship between the number of images and the angle between the

mirrors (e.g. how many images when the mirrors are angled at 45°, 60°, 90°?) and find an appropriate way to record this information.

Programmes of study: Sc4–3c

2 Cogs and pulleys Design and make a revolving stage set to aid a change of scenery. Can children change the scenery another way by using a simple pulley mechanism?

Equipment: cotton reels, elastic bands, thick card, felt tips, pulleys, string, motors, wires, batteries, wood, nails, dowel, glue, Newton meter.

Extensions: children could investigate:

- Ways to use pulleys and cogs to move other parts of scenery or equipment, e.g. curtains
- Using a Newton meter to measure the amount of force needed to lift a variety of objects from one part of the stage to another
- How to alter the speed of changing scenery
- Raising and lowering a trap door
- Moving scenery on and off in different ways.

Programmes of study: F–4b; Sc1–2b; Sc4–2f

3 Investigating puppets Investigate how puppets can move. The children then try making a moving puppet of their own.

Equipment: selection of puppets – string, glove, etc.

Extensions: children could investigate:

- Making different parts of puppets move in different ways.

Programmes of study: Sc1–1a, 1b

Machines

- Design and make machines to perform specific tasks, using Lego®, junk, motors and/or magnets
- Find out how machines have evolved over time
- Look at machines and identify component parts, e.g. cogs, gears, levers and pulleys
- Advertise own machines describing main features
- Write and tape a story about machines with sound effects
- In groups mime the actions of a real or imagined machine
- Using hydraulics

Site

- Visit a building site. Try to revisit at various stages of construction
- Draw a plan of a site considering safety, access facilities for workers
- Make a model of a building site
- Devise a questionnaire to find out what the school community would most like to see in their area
- Design and make a tower, bridge or scaffolding and test for strength and stability
- Write a letter asking for planning permission giving reasons why such a building would be beneficial to the community
- How could the site be protected during the building stage?

Buildings and builders

Structures

- Find out about buildings through the ages and around the world
- Look at the process of constructing a building, e.g. foundations first
- How are buildings affected by different climatic conditions?
- Sort and classify building materials, e.g. brick, stone, slate, copper
- Investigate the strength and suitability of different roof shapes
- Making cement

Workers

- Look at work clothing for safety. What are the main features?
- Write job descriptions indicating skill needed
- Draw up timetables and contracts. Consider problems that might arise, such as bad weather conditions, sickness or impending deadlines
- Imagine a day in the life of a particular worker
- Consider other aspects of health and safety relating to the work force

Machines activities

1 Designing machines Using Lego®, children design and make a machine to do a specific task – for instance, to lift or carry materials.

Equipment: Lego Technic®, string, dowelling, elastic bands, magnets, paper clips, Plasticine®, weights.

Extensions: children could investigate:

- Making other models using various junk material
- The strength and stability of the models made and ways of improving them
- Making models using other construction kits, e.g. Capsela®, Teko®, etc.

Programmes of study: Sc1–1a; Sc4–2f

2 Using hydraulics How can you use a

balloon to empty a truck? Children investigate the different types of load and the amount the truck can carry. Can they make the fork lift truck work? They then investigate how much the fork lift truck will lift.

Equipment: syringes, plastic tubing, model or toy tipper truck, water, balloons, balloon pump.

Extensions: children could investigate:

- What happens when you press the syringe down one unit. Children could then predict the outcome of pressing down two units, three units, etc. and test to see if they were right
- Filling the tube and syringes with water to see what difference this makes
- The relationship between the amount of air in the balloon and the height of the tipper on the truck
- Other types of pump (e.g. bicycle pump, car pump) and developing their own uses for hydraulic systems.

Programmes of study: F–1a; Sc1–2a; Sc4–2f

Building site activities

1 Scaffolding Design and make model scaffolding. Can children devise a test to find how strong it is and how stable it is?

Equipment: straws, sticky tape, pipe cleaners, string, wool, weights, slope.

Extensions: children could investigate:

- Shape to find out which is the strongest and most rigid
- Constructing scaffolding using different materials
- Scaffolding used on building sites looking at why we need scaffolding, strong shapes, platforms, safety.

Programmes of study: Sc1–1a, 1d; Sc4–2f

2 Security Investigate ways of using electricity to protect a building site.

Equipment: bulbs, buzzers, batteries, wires, bulb-holders, screw drivers, silver foil, drawing pins, paper clips, wood, circuit breakers, card, clothes peg.

Extensions: children could investigate:

- Making an alarm that is set off by a pressure pad
- Ways of making a series of lights flash on and off
- Materials that conduct electricity and therefore could be used in making switches.

Programmes of study: Sc3–1c; Sc4–1b

3 Visiting a site Visit a building site. Ask the children to make a list of all the different materials which they see. When you get back to school they should try to sort the list into three or four sets so that the materials have something in common.

Equipment: paper.

Extensions: children could investigate:

- The reasons for choosing particular sets
- Identifying where certain materials are used and why
- Sorting materials into solids, liquids and gases
- Listing all the materials they can see in the classroom into sets, e.g. natural/made.

Programmes of study: Sc3–1a, 1e

Structures activities

1 Roofing Roofs are designed to keep out rain. Children should design and make a roof which will do this. Consider some of the following: Why do we need rafters?

Why do we use felt? Why do we need a gutter? The roof will be tested – it will have to keep a paper towel dry.

Equipment: balsa wood, polythene, card, felt, scissors, saw, glue.

Extensions: children could investigate:

- How to make the test fair
- Listing their roof's faults after testing and trying to improve on their design
- Roofs in other parts of the world, e.g. where there is little rain.

Programmes of study: F–1a; Sc1–1d

2 Making cement Devise an investigation to find out which is the 'best' consistency for each of the mixtures.

Equipment: Polyfilla®, cement, plaster of Paris, timer, water, containers.

Extensions: children could investigate:

- The changes that take place when the powders are mixed together with varying amounts of water and record their observations
- The uses for each of the white powders in everyday life
- Making bricks using clay and observe the changes as the clay dries out.

Programmes of study: Sc1–1a, 1d; Sc3–2b, 3b

Activities about workers

1 Work clothing   Consider the clothes worn by workers on a building site. Think about safety and wearability. Devise some investigations to test materials to see if they are waterproof, safe and hard wearing.

Equipment: as required by the children, but provide a variety of fabrics, hard polythene, sand paper for rubbing tests, masses for strength tests.

Programmes of study: Sc1–1d; Sc3–1a

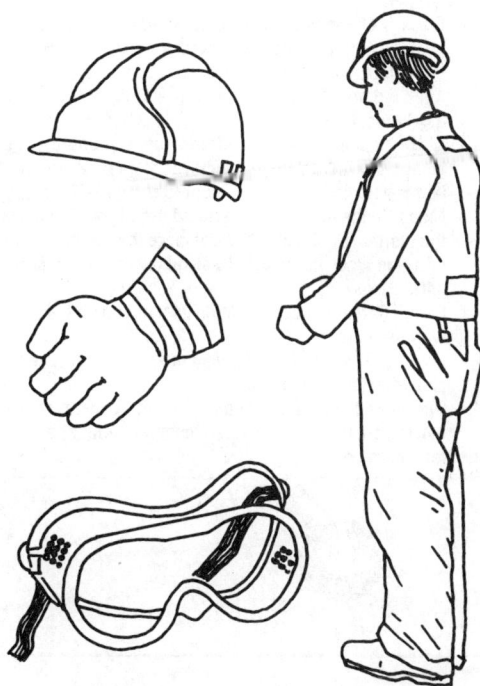

Energy

Ourselves and our world

- Food – find out about the energy content of the food we eat; find the calories/kilojoules on packages; identify carbohydrates, proteins, fats
- Muscle power: pushes and pulls
- Design a fair test to find 'Muscle person'
- Temperature: measure the temperature under your arm, under your big toe and compare
- Can you think why you wear a hat in winter?
- Keeping warm: look at the colours and textures of materials worn in different climates
- Investigate keeping jars of water warm/cold using different materials by regularly measuring the temperature
- Classify changes as permanent or temporary when everyday ingredients are heated or cooled, e.g. wax, chocolate, water, butter
- Keep a diary of household energy sources used daily, e.g. kettle, toaster
- Make a collection of the different types of energy, e.g. solar powered calculator, pictures of gas, electricity, oil, water powered objects, nuclear fuel
- Measure the temperature around the school. Which is the warmest/coldest area? Compare the temperatures at different times of the day. Can you give reasons for these differences?
- Compare the fuel consumption in school with the weather data
- Investigate ways of conserving energy, e.g. draught excluders, aluminium foil
- Guess the temperature of a number of everyday foods, e.g. ice cream, cup of tea, and compare with actual temperature

On the move

- Jet propulsion: compare the distance the balloon travels along a string, with the amount of air
- Can you make a wheeled vehicle move using balloon power
- Hydraulics: can you make the fork lift truck work using the syringes? How can balloon power help to lift things?
- Hovercraft: using a polystyrene tile and a balloon, how far can you make it travel?
- Design and make a water wheel
- How many ways can you make your boat move, e.g. paddle boat using elastic band power, straws?
- Can you alarm the lift, so that when it hits the floor it makes a noise to show it is ready to go up again?

Controlling energy

- Elastic band powered models, e.g. crawlers, mangonels, mirror target games. Investigate how you can make your models go farther
- Power your vehicles using an electric motor. Make use of gears, belts and levers. Make your Lego® model move. Make lights for your model
- Investigating toys with different energy sources; how far, from one turn, will it travel?
- Investigate toys which spin
- Design and make a spinning disc. Which is more efficient – wool, string, nylon, thread?
- Design and make a rocket launcher. How far will your rocket travel? Can you use it to launch something along the ground?
- Design and make a lift, using junk material and an electric motor

Ourselves and our world activities

1 Cooling down Can you devise some ways of cooling down hot milk so that it is safe for a baby to drink?

Equipment: hot water, containers, thermometers, cold water, metal spoon, materials, baby bottles, timer.

Extensions: children could investigate:

- Making predictions as to which will cool down first
- Keeping an ice cube for as long as possible
- Other materials (e.g. wax, chocolate, butter, water) and the effects of heating or cooling on them.

Programmes of study: Sc1–1b, 1d; Sc3–1b, 2b, 2f

2 Heating in buildings Measure the air temperature in different parts of the room and compare them. What is the average temperature? Find another room in the

90

building and measure the air temperature in different places. Compare the average temperature with the first room.

Equipment: thermometers, paper.

Extensions: children could investigate:

- The effects that sun can have on room temperature
- How electricity/gas is metered and look at bills to see the costing
- Ways of conserving energy.

Programmes of study: F–2a, 2b, 2c, 2d, 4b

3 World energy sources Ask the children to consider all the ways in which they use energy daily at home and in school. They then identify the energy source – gas, electricity.

Equipment: reference books and materials produced by Powergen, British Gas, Nuclear Electric.

Extensions: children could investigate:

- How much electricity is used by certain appliances
- Ways of conserving energy in the home, at school, and so on
- Solar power and alternative forms.

Programmes of study: F–2c, 4a

On the move activities

1 Jet machine Design and make a balloon-powered device which will travel along a line as quickly as possible.

Equipment: balloons, sticky tape, drinking straws, paper, fishing line, card.

Extensions: children could investigate:

- How far the device will travel with different amounts of air in the balloon
- Different shaped balloons.

Programmes of study: Sc1–1d; Sc4–2h

2 Boats How many ways can children make a boat move?

Equipment: boats, junk materials, balsa wood, elastic bands, nails, electric motor, bellows, fan, balloons, card.

Extensions: children could investigate:

- The relationship between the number of turns the paddle makes to the distance travelled
- Different boat shapes to find the fastest
- The quickest way of moving the boat from A to B.

Programmes of study: Sc1–1a, 1b, 1d; Sc4–2g

3 Sound energy Ask the children to see if sound travels through solids. Provide a ticking clock and a metre rule. First the children find the optimum distance at which they can hear the clock. They then place the rule so that it is touching the clock and listen to see if the ticking sounds the same.

Equipment: clock, metre rule, water, bell, tuning fork, ping pong balls, access to running water

Extensions: children could investigate:

- Testing the bell in water (varying the amounts of water)
- Observing what happens when the tuning fork is placed near running water and close to other materials.

Programmes of study: Sc4–3c, 3g

Controlling energy activities

1 Storing energy Children design and make a mangonel. What difference does the size or weight of the missile make to the firing distance?

Equipment: wood, plastic spoons, rubber bands, long nails, screw eyelet, peg, Plasticine® and other kinds of loads.

Extensions: children could investigate:

- The difference made by changing the shape and mass of the load
- How firing power can be improved by trying thicker elastic, more turns, a different spoon, changing the position of the spoon or the firing release position.

Programmes of study: Sc1–1a, 1d, 2b, 2c, 3b; Sc4–2h

2 Going up Children design and make a working lift. Can they alarm the lift, so that when it reaches ground level it makes a noise to show that it is ready to go up again?

Equipment: junk, electric motors, string, cotton reels, dowel, sticky tape, buzzers, silver foil, wires, drawing pins, paper clips, peg.

Extensions: children could investigate:

- How much load the lift can carry
- Ways of controlling the speed of the lift
- The electrical circuits and try adding lights.

Programmes of study: Sc4–1b, 1c

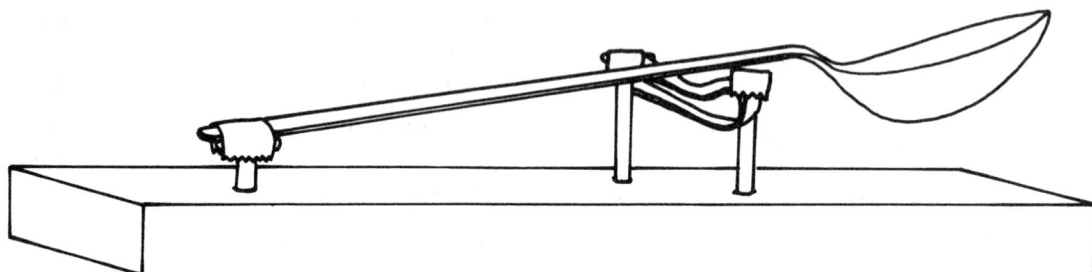

Our world and beyond

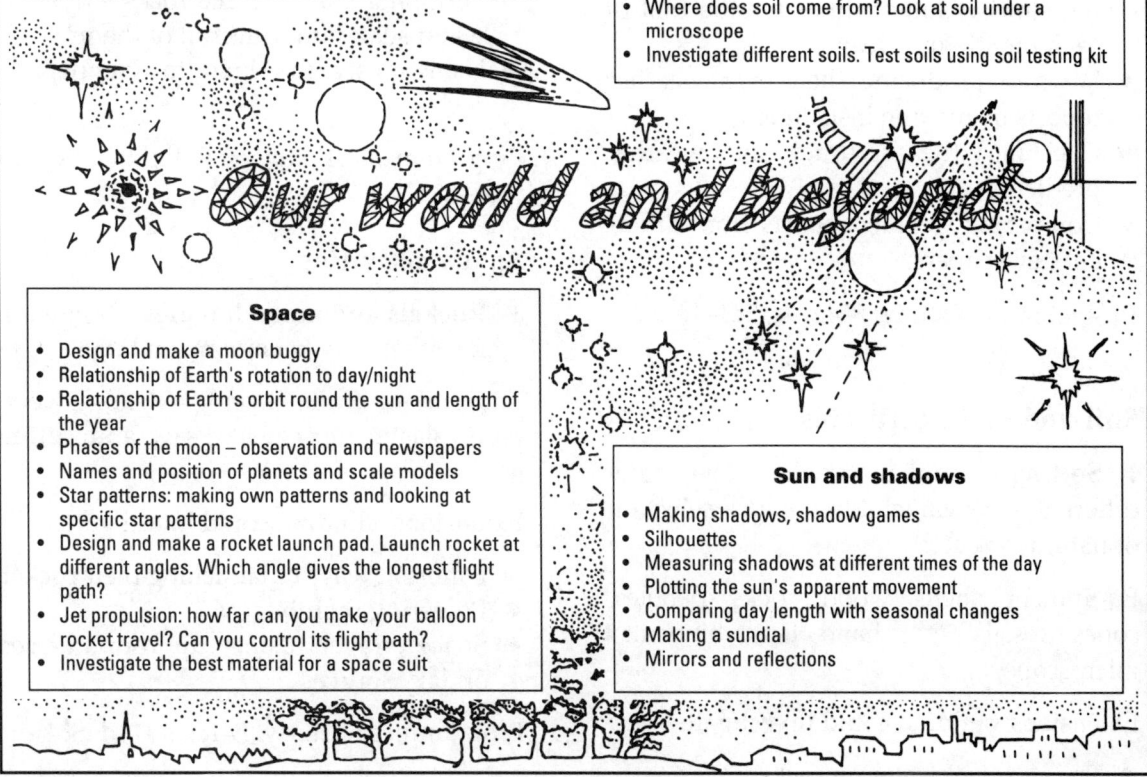

Weather

- Observing the weather over a period of time
- Designing a weather chart
- Designing own weather symbols and own weather report
- Design, make and use their own weather instruments/rain gauge
- Collecting weather data from the newspapers

Soil and rocks

- Sorting natural materials: shells, pebbles, sand
- Looking at building materials, e.g. bricks, stones, tiles. How waterproof are these building materials?
- Rubbings and observational drawings of bricks, stone
- Looking at different types of rock using a hand lens and noting its features. Devise a test to see which is the hardest
- Where does soil come from? Look at soil under a microscope
- Investigate different soils. Test soils using soil testing kit

Space

- Design and make a moon buggy
- Relationship of Earth's rotation to day/night
- Relationship of Earth's orbit round the sun and length of the year
- Phases of the moon – observation and newspapers
- Names and position of planets and scale models
- Star patterns: making own patterns and looking at specific star patterns
- Design and make a rocket launch pad. Launch rocket at different angles. Which angle gives the longest flight path?
- Jet propulsion: how far can you make your balloon rocket travel? Can you control its flight path?
- Investigate the best material for a space suit

Sun and shadows

- Making shadows, shadow games
- Silhouettes
- Measuring shadows at different times of the day
- Plotting the sun's apparent movement
- Comparing day length with seasonal changes
- Making a sundial
- Mirrors and reflections

Weather activities

1 Measuring the wind Design a simple instrument to measure the force of the wind.

Equipment: battery powered fan, hair drier (to be used only under adult supervision), tissue paper, tights, straws, dowel, card, papers, wire.

Extensions: children could investigate:

- Ways of calibrating their instrument for force and direction
- Weather reports to see if the wind speed is reported
- Making and testing other measuring instruments – rain gauges, etc.

Programmes of study: F–4b; Sc1–2a, 2b, 2c

2 Weather records Using the weather charts provided, guess what part of the

93

year is being reported. Where in the world was the average temperature highest? Using the charts, estimate the sunrise/sunset, moonrise/moonset for the following Monday.

Equipment: weather charts for one week cut from newspapers, paper, an almanac.

Extensions: children could investigate:

- Using the almanac to compile a graph of the sunrise/sunset times for the first of each month for a year
- Ways of producing their own weather reports by making instruments
- Checking the accuracy of weather predictions
- Comparing weather reports from different newspapers.

Programmes of study: F–4a, 4c; Sc1–1e

Soil and rocks activities

1 Sorting natural materials How many different ways can children sort a selection of natural objects?

Equipment: shells, wood, rocks, pebbles, cones, fossils, etc., hand lens, binocular microscope.

Extensions: children could investigate:

- The effects of water on rocks
- Different ways of recording, for example using keys
- Sorting the objects and asking other children to guess the criteria used
- The size and shape of various particles in the rocks using a microscope and draw them
- Using secondary sources to find out more information about the rocks.

Programmes of study: F–1c; Sc3–1d

2 Investigating rocks Children devise their own test to see which rock is the hardest.

Equipment: rocks, scissors, sandpaper, needles, coin weights, cardboard tubes, droppers, water.

Extensions: children could investigate:

- Which implements could be used for a scratch test
- Dropping water on the rocks – does the water soak into some/all of them?
- Using research skills to identify the rocks.

Programmes of study: F–1c; Sc1–1a, 1d; Sc3–1a

Space activities

1 Rockets and moon buggies Design and make either a rocket or a moon buggy.

Equipment: junk, Lego®, wheels, wood, glue, plastic tubing, hacksaw and cutting jig.

Extensions: children could investigate:

- Different ways of launching their rocket
- Powering their buggy
- Space travel using appropriate secondary sources.

Programmes of study: F–1c; Sc1–1d; Sc4–2f

2 Investigating the solar system Using appropriate secondary sources, the children find out about the relative positions of the Earth, the sun and the moon. They should observe the sky at night and using appropriate charts investigate the phases of the moon.

Programmes of study: Sc4–4c, 4d

3 Investigating the stars Children make their own star pattern and invent a name for it.

Equipment: star charts, card tubes, black paper, tissue paper, chalk.

Extensions: children could investigate:

- Looking at the stars in the night sky and noticing the difference in brightness
- The names and their origins of the stellar constellations
- Plotting the apparent 'movement' of the sun.

Programmes of study: Sc4–4b

Sun and shadows activities

1 Investigating silhouettes Make a silhouette.

Equipment: overhead projector, paper, chalk, scissors, glue.

Extensions: children could investigate:

- Making shadows larger
- Measuring shadows at different times of the day
- Plotting the sun's apparent movement.

Programmes of study: Sc4–3b, 4b

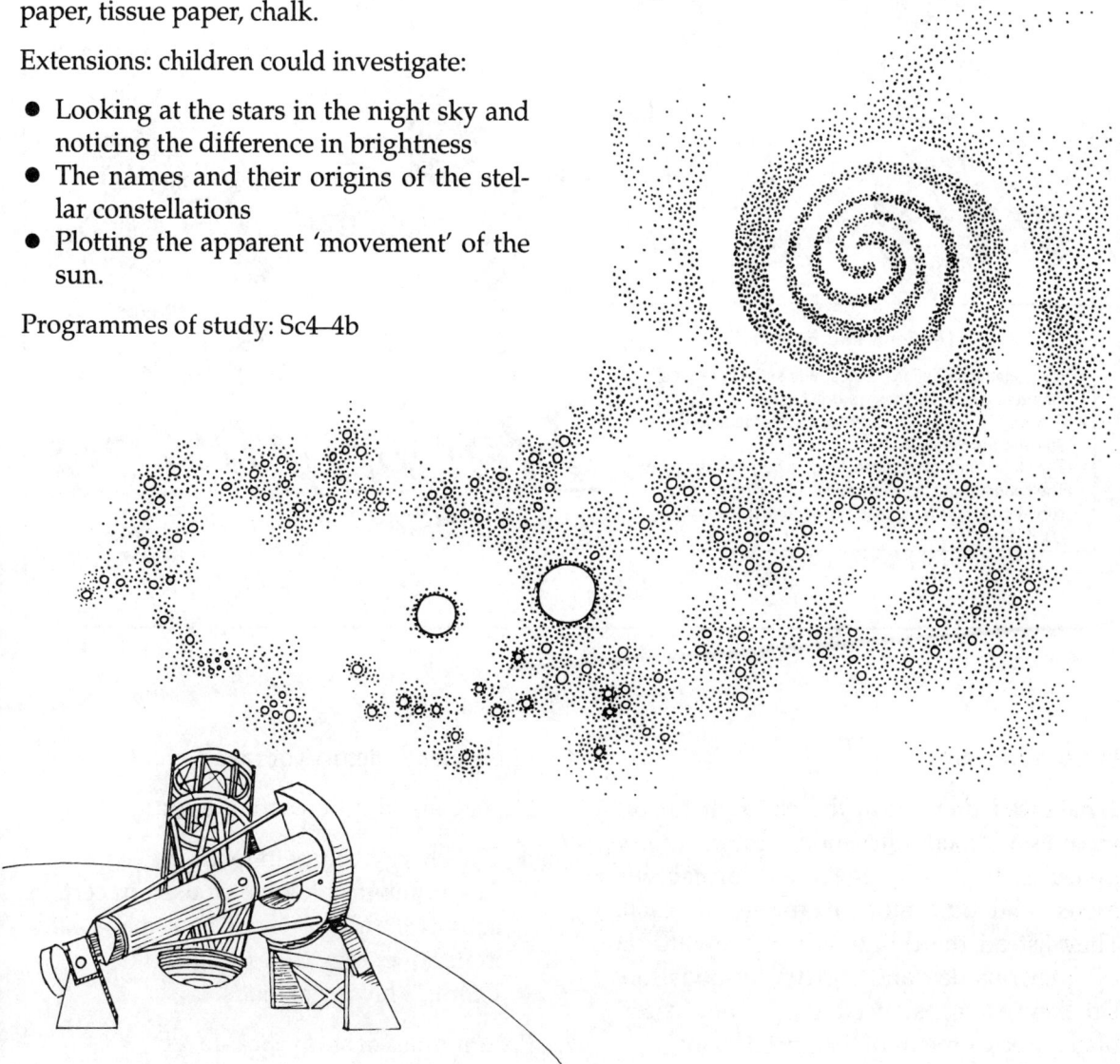

Food

- Healthy eating, the need for a balanced diet, energy values, kilojoules
- Investigating food eaten by other cultures, devising balanced meals
- How does yeast work? Making bread, investigating yeast mixtures, investigating different breads
- Investigating food chains
- Looking at changes in foods: cooking and decay
- Investigating water in food

The environment

- Water – the need for clean drinking water, identifying samples of water, how our water is made clean.
- Cleaning up a river, the effects of pullution
- The water cycle
- Plants or flowers – growing seeds under different conditions, making fair tests
- Investigating flowers, pollination
- Identifying types of plants
- Investigating leaves: identifying where leaves get dirtiest
- Minibeasts – pond dipping, minibeast hunts, life cycles of creatures, food chains
- Investigating rubbish and decay
- Improving an area
- Investigating soil, rocks and fossils. Investigating fossils

Hygiene and health

- Investigating variation – similarities and differences, inherited characteristics – stereotyping
- Growing up, life cycles, pregnancy, reproduction in animals, growing old
- The importance of exercise, lung power, breathing rate
- The need for oral hygiene, using disclosing tablets, surveys on visits to dentists, orthodontists

Healthy living

Food activities

1 Balanced diets Ask the children to sort pictures of food into food groups (dairy products, fruit and vegetables, farinaceous foods) and add more examples to each. They list all the things which they ate on the previous day and sort those foods too. Do they eat a balanced diet? They could plan a week's menu of 'balanced foods'.

Equipment: pictures of foods, charts.

Extensions. children could investigate:

- The energy values in foods
- The number of kilojoules used in certain activities and work out how many kilojoules of energy they use in a day
- People's favourite foods.

Programmes of study: Sc2–1a, 2b

2 Water in food Ask the children to find a way of measuring the amount of water in spaghetti. Ask them to try to make their test fair by weighing the spaghetti carefully before and after cooking.

Equipment: dried foods (e.g. various types of pasta, rice, pulses), water, access to a cooker, balance.

Extensions: children could investigate:

- How much water there is in other foods, such as bread, by drying out the foods in an oven
- Other ways of preserving foods
- Preparing a three course meal without using water (or a microwave oven)
- Decaying food.

Programmes of study: Sc2–5e

3 Investigating yeast Ask the children to add some yeast and sugar to warm water, place the mixture in a test tube or bottle with a balloon over the neck, leave for ten minutes in a warm place and note any changes. They can then try different mixtures – water and yeast, sugar and water, sugar, yeast and water – and observe what happens.

Equipment: yeast, ingredients and recipes for bread making, a variety of leavened, unleavened bread.

Extensions: children could investigate:

- Observing and classifying breads in different ways
- Making breads.

Programmes of study: Sc1–1d; Sc3–2b

Hygiene and health activities

1 Breathing The children observe the number of breaths a partner takes in a minute. This is repeated after a minute's exercise and the results recorded. They observe what happens when a deep breath is taken, measuring chest size, etc. They then investigate lung capacity.

Equipment: timer, tape measure, plastic tank or plastic bottle, tubing etc.

Extensions: children could investigate:

- Using secondary sources to find out what happens when we breathe, how oxygen gets into the blood, how the air we breathe out differs from the air we breathe in
- The effects of smoking.

Programmes of study: F–1c; Sc2–2c, 2d, 2h

2 Variation and inheritance Look at a photograph of the class. Identify the similarities and differences within the group. Discuss and investigate other similarities and differences as well as the obvious ones. Remind the children about characteristics, such as attached/detached earlobes, tongue rolling ability. Encourage the children to record their findings in a variety of ways.

Extensions: children could investigate:

- 'Stereotyping'
- Inherited characteristics
- Human reproduction.

Programmes of study: Sc2–1a, 4a

Healthy environment activities

1 Clean water The children test samples of water and observe them using a microscope. See if there is anything dissolved in the water by leaving some in a warm place for a day or two to dry up, or see if the water helps things to float. The children should write down their observations. Say which, if any, of the samples would be safe to drink but, of course, children shouldn't try them!

Equipment: water samples (rainwater, sea water, river/stream water, drinking water, water and sugar), microscope, filter paper, funnels, sand, gravel, soil.

Extensions: children could investigate:

- Filtering a mixture of soil and water
- Using secondary sources to investigate how drinking water is purified
- The amounts of water used and its purposes
- The water cycle
- Desalination of water
- Rainfall figures
- Microbes in water
- The effect of polluted water
- Solubility of substances.

Programmes of study: Sc1–1a; Sc2–5e; Sc3–2e, 3b, 3c, 3d

2 Investigating plants The children grow plants under different conditions and observe them through their life cycle, discussing how to make the test fair and what to measure.

Equipment: seeds, plants, containers, different mediums.

Extensions: children could investigate:

- Growing vegetables
- Plants which grow in walls, pavements, and so on
- Pollination
- Leaves
- Designing and making plant waterers.

Programmes of study: Sc2–1b, 3a, 3c, 3d, 5a

3 Investigating pond life The children investigate pond life using relevant secondary sources. They identify animals and water plants after pond-dipping, then make sketches of the pond, identifying where the various creatures and plants were found.

Equipment: pond-dipping equipment, plastic tanks, relevant reference materials.

Extensions: children could investigate:

- The life cycle of a pond creature
- How rivers are cleaned
- Drawing up plans for a school pond
- Listing all local ponds, rivers, and so on
- The food chains of aquatic creatures.

Programmes of study: F–1a; Sc1–1b, 3d, 3e; Sc2–1a, 5c, 5d

9

SOURCES OF INFORMATION

Department for Education (1995), *Science in the National Curriculum*

School Curriculum and Assessment Authority (1995), *Planning The Curriculum at Key Stages 1 and 2*

School Curriculum and Assessment Authority (1995), *Consistency in Teacher Assessment: Key Stages 1 to 3*

School Curriculum and Assessment Authority (1995), *Exemplification of Standards* (English, Mathematics and Science available)

School Curriculum and Assessment Authority (1995), *Key Stages 1 and 2 – Assessment Arrangements*

Cheshire County Council Education Services (1995), *Science Key Stages 1 and 2 – An Aid to Planning*

Cheshire County Council Education Services (1995), *Science Key Stages 1 and 2 – Indicators of Progression*

10

PHOTOCOPIABLE SHEETS

On the following pages photocopy masters are included to enable you to utilise some of the charts used as examples throughout this book. Copy them at 130% magnification to fit an A4 sheet.

Science class record sheet (Key stage 1)

Class .... to .... 19............

Foundation Science	1	a	b	c	d			
	2	a	b	c				
	3	a						
	4	a	b					
	5	a	b					
Science 1 **Experimental and investigative science**	1	a	b	c				
	2	a	b	c				
	3	a	b	c	d	e	f	
Science 2 **Life processes and living things**	1	a	b					
	2	a	b	c	d	e	f	
	3	a	b	c				
	4	a	b					
	5	a	b					
Science 3 **Materials and their properties**	1	a	b	c	d	e	
	2	a	b					
Science 4 **Physical processes**	1	a	b	c				
	2	a	b	c	d			
	3	a	b	c	d	e		

Topics covered

Year group	Autumn		Spring		Summer		Teacher
Nursery							
R							
1							
2							

Science class record sheet (Key stage 2)

Class .... to .... 19.............

Foundation Science	1	a	b	c	d				
	2	a	b	c	d				
	3	a	b						
	4	a	b	c					
	5	a	b						
Science 1 **Experimental and investigative science**	1	a	b	c	d	e			
	2	a	b	c					
	3	a	b	c	d	e			
Science 2 **Life processes and living things**	1	a	b						
	2	a	b	c	d	e	f	g	h
	3	a	b	c	d				
	4	a							
	5	a	b	c	d	e			
Science 3 **Materials and their properties**	1	a	b	c	d	e			
	2	a	b	c	d	e	f		
	3	a	b	c	d	e			
Science 4 **Physical processes**	1	a	b	c	d				
	2	a	b	c	d	e	f	g	h
	3	a	b	c	d	e	f	g	
	4	a	b	c	d				

Topics covered

Year group	Autumn		Spring		Summer		Teacher
3							
4							
5							
6							

Weekly planning and assessment sheet

.... to 1996

Subject	Lesson	Objective	Assessment	Future planning
English				
Maths				
Science				

Weekly planning and assessment sheet

.... to 1996

Subject	Lesson	Objective	Assessment	Future planning
History				
Geography				
Art				
Music				
RE				
Technology				
IT				
PE				
Notes				

Topic Planning Sheet

	Autumn	Spring	Summer
R			
Y1			
Y2			
Y3			
Y4			
Y5			
Y6			

Topic planning and assessment sheet

Area of study/topic.. **Class**..............................

Term......................

Science	Learning objective (from PoS)	Activities	Method of recording	Specific resources	Assessment (O, P, Q, S)	E
Foundation						
Science 1						
Science 2						
Science 3						
Science 4						